Suresh Sundaramurthy · Surendra Pratap Singh ·
Anuj Kumar Verma · Mohammad Jawaid ·
Eldon R. Rene
Editors

Black Liquor Waste Management in the Pulp and Paper Industry

Springer

Editors
Suresh Sundaramurthy
Maulana Azad National Institute of Technology
Bhopal, Madhya Pradesh, India

Anuj Kumar Verma
Akemi Minerals India Pvt. Ltd.
Kachchh, Gujarat, India

Eldon R. Rene
IHE Delft Institute for Water Education
Delft, The Netherlands

Surendra Pratap Singh
Pulp and Paper Research Institute
Jaykaypur, Odisha, India

Mohammad Jawaid
Department of Chemical and Petroleum Engineering
College of Engineering
United Arab Emirates University
Al Ain, Abu Dhabi, United Arab Emirates

ISSN 2731-0426 ISSN 2731-0434 (electronic)
Sustainable Materials and Technology
ISBN 978-981-96-4236-6 ISBN 978-981-96-4237-3 (eBook)
https://doi.org/10.1007/978-981-96-4237-3

This Springer imprint is published by the registered company Springer Nature Singapore Pte Ltd.
The registered company address is: 152 Beach Road, #21-01/04 Gateway East, Singapore 189721, Singapore

Sustainable Materials and Technology

Series Editors

Mohammad Jawaid, Chemical and Petroleum Engineering, United Arab Emirates University, Al Ain, UAE

Anish Khan, Centre of Excellence for Advanced Materials, King Abdulaziz University, Jeddah, Saudi Arabia

Sustainable Materials and Technology (SMT) book series publishes research monographs (both edited and authored volumes) showcasing the latest developments in the field and comprehensively covering topics such as:

- Recycling of waste into useful material and their energy applications
- Catalytic action of Nano oxides for efficient carbon reforming process
- Sustainable technologies for plastic transformation
- Bifunctional nanoparticles for sustainable water splitting applications
- Sustainable dying and printing
- New materials from waste
- Sustainable Manure Management and Technology: Potentials, Uses and limitations
- Sustainable Mechanical Engineering Approach
- Sustainable biochemistry for the improvement of health
- Sustainable development of Mechanical recycling of automotive components
- Sustainable-waste recycling and conversion in useful materials for different applications
- Sustainable development of inexpensive Nano-photo catalysts
- Sustainable development of recycling of discarded lithium ion batteries
- Modern sustainable cement and concrete
- Sustainable adsorbent for hazardous removal
- Sustainable superior electromagnetic shielding materials
- Excellent sustainable nanostructured materials for energy storage device
- Sustainable development of heavy metal detoxification from water
- Carbon dioxide utilization for sustainable energy
- Sustainable development in green syntheses of materials
- Environment friendly and sustainable cloth for garments application
- Sustainable design and application of eco-materials
- Nanoparticles for sustainable environment applications
- Sustainable remediation of industrial contaminated water towards potential industrial applications
- Biomaterials for sustainable bioremediations

Foreword

A wide variety of biomass is available all around the world. Most of the biomass exists as a by-product from manufacturing industries. Pulp and paper mills contribute to a higher amount of these biomasses mostly discharged as waste or otherwise could be explored as potential source of industrially useful by-products. Biomasses from other manufacturing industries/sources have been used to produce different kinds and grades of biomaterials such as those used in industrial and medical applications.

Black liquor is the alkaline wastewater produced during the chemical pulping process in paper making industry. Black liquor contributes to 10%–15% of the total pulp and paper industry waste by volume but almost 90% of total pollution as chemical oxygen demand. Black liquor is typically concentrated and combusted in alkali recovery boilers to produce energy and reclaim chemicals. Black liquor could be used to produce many renewable materials for industrial applications and could replace some fuels and commercial materials. This book on Black Liquor Management in the Pulp and Paper Industry discusses the various options for waste to wealth through a circular economy framework.

This book consists of 10 chapters that highlights different objectives of black liquor waste utilization, characterization as well as circular economy and green opportunities for the pulp and paper industry.

In the first chapter "Black Liquor Chemical, Physical and Combustion Properties and Their Importance in Chemical Recovery Operations," the physical, chemical, rheological, combustion, and thermal properties of black liquor and their importance in chemical recovery operations have been given in detail. The properties of black liquor covered in the chapter include pH, total solids, total suspended solids, residual active alkali, total alkali, ratio of organic to inorganic compounds, elemental composition of black liquor solids, colloidal stability, non-process elements and their Impact on recovery operations, impact of chlorides and potassium on chemical recovery systems, impact of silica and other NPEs, viscosity of black liquor, boiling point elevation, fouling tendency, surface tension, thermal and combustion properties, calorific value, and swelling volume ratio. In the second chapter "Potential of Synthesis of Polyurethane Foam from Black Liquor," the potential use of black liquor as lignin source for synthesis of polyurethane foam has been explained. Lignin has found its diverse applications and one of the potential utility studied for making polyurethane foam. Polyurethane is a versatile polymer that belongs to the class of synthetic polymers known as polyurethanes. It is formed through the reaction of polyols and di-isocyanates. The reaction between these two components results in the formation of urethane linkages, which are responsible for the polymer structure.

In the pulp and paper industry, lignin is present in black liquor and there is need of extraction, purification, and characterization of the lignin to explore its application in various industries. The quantitative and qualitative analytical methods of lignin is required to explore its various application in industries. In the scope of the third chapter "Trends in Analytical Techniques for Lignin Quantification and Characterization," the details of analytical techniques of lignin and its significance has been mentioned including the characterization of chemical structure, particle size, and molecule size determination.

In the fourth chapter "Modification of Black Liquor Properties to Enhance the Chemical Recovery Efficiency," the modification of black liquor properties to enhance chemical recovery efficiency has been mentioned.

In the scope of the fifth chapter "Development of Modern Materials from Lignosulphonates," the possibility of developing modern materials from lignosulphonates has been explained including construction, mineral powder, chemical industry, textile industry, metallurgical industry, dust control, lubricants, antioxidants, fire retardants, automotive interiors, petroleum industry, ceramics, fire retardant, rubber vulcanization, food for animal species, chelating agents, binders, emulsifiers, dispersants, sodium-ion batteries, cathodes for lithium-ion batteries, and binders for lithium-ion batteries.

In the sixth chapter "Recent Advances in Chemical Recovery Furnace for the Efficient Combustion of Black Liquor," the recent advances in chemical recovery by efficient combustion of black liquor in furnace has been covered. In recent times, the recovery furnace is not only used to recover the cooking chemicals in paper industry but also to generate the energy in terms of electricity. Black liquor contains several organic, inorganic, and non-process elements and hence it is very essential to control the corrosion in the furnace by increasing the heat transfer rate.

The seventh chapter "Production of Bioethanol as a By-Product in the Pulp and Paper Industry" covered the production of bioethanol as a by-product in the and paper industry. The chapter delves into the innovative approach of utilizing lignocellulosic biomass, the primary raw material in paper production, as feedstock for bioethanol generation. This chapter provides a comprehensive overview on the challenges faced towards making lignocellulosic bioethanol into commercially viable reality.

The eighth chapter "Valorization of Rice Straw for Ethanol Production and Lignin Recovery Through Acidic and Alkaline Pre-treatments" described valorization of rice straw for ethanol production and lignin recovery through acidic and alkaline pretreatments. The benefits of pretreatment of lignocellulosic materials for the production of ethanol are highlighted in this chapter, along with recommendations for the recovery of glucan and xylan from rice straw following pretreatments. However, all lignocellulose biopolymers should be recovered in a single step in order to obtain a more sustainable bio-refinery-based strategy.

In the ninth chapter "Facts on Climatic Effect of Waste Paper Recycling in the Pulp and Paper Industry," the limited climate benefits of global recycling of pulp and paper have been explained by making review of studies conducted on aspect of recycling of paper which is said to decrease the cutting of trees and deforestation and increase ecological aspects.

The tenth chapter "Opportunities for Green Technology Initiatives in the Indian Paper Industry" covered the opportunities for green technology initiatives in the Indian paper industry. The growth of the paper industry is intricately linked with the booming knowledge and commercial economy of India. These produce higher quantities of effluents because they consume a lot of raw materials, chemicals, energy, and water. It is noteworthy that this industry is 1 of the 17 that the Central Pollution Control Board has classified as extremely polluting and having insufficient adherence to environmental regulations.

The editors of this book have shown that, the black liquor waste management, besides their conventional use, can be converted into high value for versatile applications. It is revealed that the black liquor generated from the mills provides ingredients for the synthesis of polyurethane foam, modern materials, chemical recovery, and production of bioethanol. I am sure readers can easily understand the each chapter's goal, advancement, and specific applications.

New Delhi, India

Dr. Rakesh Kumar Jain
Technical Expert
United Nations Industrial Development
Organisations (UNIDO)
Former Director–CPPRI and KNHPI

Preface

Black liquor is the alkaline wastewater generated during the chemical pulping process in the paper manufacturing sector. It accounts for 10%–15% of the total waste produced by the pulp and paper industry. The management of black liquor, including the recovery of pulp, clean, and value-added chemicals, provides valuable insights into the scientific and technical advancements necessary for the growth of the pulping industry, as well as for national economic development and environmental sustainability, particularly in light of increasing concerns regarding climate change.

This book consists of 10 chapters that highlights different black liquor waste management in the pulp and paper industry objectives. This book highlights latest developments in pulping process from different sources of raw materials, properties, and recovery technologies of black liquor, extraction of value-added chemicals, and green technology initiatives in Indian paper industry.

In first chapter "Black Liquor Chemical, Physical and Combustion Properties and Their Importance in Chemical Recovery Operations", the physical, chemical, rheological, and combustion and thermal properties of black liquor and their importance in chemical recovery operations have been given in detail. In second chapter "Potential of Synthesis of Polyurethane Foam from Black Liquor", elemental composition of black liquor solids, colloidal stability, non-process elements and their impact on recovery operations, potential use of black liquor as lignin source for synthesis of polyurethane foam has been explained.

In the scope of third chapter "Trends in Analytical Techniques for Lignin Quantification and Characterization", the details of analytical techniques of lignin and its significance has been mentioned including characterization of chemical structure, particle size, and molecular size determination. In fourth chapter "Modification of Black Liquor Properties to Enhance the Chemical Recovery Efficiency", the modification of black liquor properties to enhance chemical recovery efficiency has been mentioned. In the scope of fifth chapter "Development of Modern Materials from Lignosulphonates", the possibility of modern materials from lignosulphonates has been explained including construction, mineral powder, chemical industry, textile industry, metallurgical industry, dust control, lubricants, antioxidants, fire retardants, automotive interiors, petroleum industry, ceramics, fire retardant, rubber vulcanization, food for animal species, chelating agents, binders, emulsifiers, dispersants, sodium ion batteries, cathodes for lithium ion batteries, and binders for lithium ion batteries.

In sixth chapter "Recent Advances in Chemical Recovery Furnace for the Efficient Combustion of Black Liquor", the recent advances in chemical recovery by efficient combustion of black liquor in furnace has been covered. In the recent times, the recovery furnace is not only used to recover the cooking chemicals in paper industry but also generate the energy in terms of electricity. The seventh chapter "Production of Bioethanol as a By-Product in the Pulp and Paper Industry" provides a comprehensive overview on the challenges face towards making lignocellulosic bioethanol into commercial viable reality.

In eighth chapter "Valorization of Rice Straw for Ethanol Production and Lignin Recovery Through Acidic and Alkaline Pre-treatments" described the valorization of rice straw for ethanol production and lignin recovery through acidic and alkaline pre-treatments. The benefits of pretreatment of lignocellulic materials for the production of ethanol are highlighted in this chapter, along with recommendations for the recovery of glucan and xylan from rice straw following pretreatments.

In ninth chapter "Facts on Climatic Effect of Waste Paper Recycling in the Pulp and Paper Industry", the limited climate benefits of global recycling of pulp and paper has been explained by making review of studies conducted on aspect of recycling of paper which is said to decrease the cutting of trees and deforestation and increase ecological aspects. The tenth chapter "Opportunities for Green Technology Initiatives in the Indian Paper Industry" covered the opportunities for green technology initiatives in Indian paper industry. The growth of the paper industry is intricately linked with the booming knowledge and commercial economy of India.

We sincerely hope that our contribution throughout this book will be a valuable asset to researchers, instructors, decision-makers, practicing professionals, senior undergraduate and graduate students, and others interested in black liquor waste management in the pulp and paper industry. The book could even be used as a reference book for researchers teaching courses dealing with waste valorization. All the chapters were contributed by professionals from academia and government laboratories from various countries. The editors thank the Springer for believing in our book project. We gratefully acknowledge all the authors who have contributed to this book for sharing their views and research findings with the scientific and professional communities through our book. The views or opinions expressed in each chapter of this book are those of the authors.

Bhopal, India — Prof. Suresh Sundaramurthy
Jaykaypur, India — Dr. Surendra Pratap Singh
Kachchh, India — Dr. Anuj Kumar Verma
Al Ain, UAE — Prof. Mohammad Jawaid
Delft, The Netherlands — Dr. Eldon R. Rene

Contents

About the Editors

Prof. Dr. Suresh Sundaramurthy is a Faculty and Head of Department, Chemical Engineering, Maulana Azad National Institute of Technology (MA-NIT) Bhopal, India. He has received Ph.D. in Chemical Engineering from Indian Institute of Technology Roorkee, India. Dr. Suresh was the recipient of Post doctoral Fellow awarded by IUSSTF, Govt. of India, and that time, he was associated at the City University of New York, USA. He was also the recipient of Visiting Faculty and Researcher awarded by Govt. of India. During this award, he was deputed to the Asian Institute of Technology Thailand and International Centre for Materials Science (JNCASR), Bengaluru, India. He has published over 150 journal articles in the area of new sorbents and catalysts, reactor design, environmental biotechnology, waste-to-energy, nanoengineered materials, gas separation, sensing of toxic gases, process safety, decarbonization and green hydrogen.

Dr. Surendra Pratap Singh is currently working as the Department Head (Quality Control and Labs) in Harihar Polyfibers, GRASIM Industries Ltd. Kumarapatnam, Haveri, Karnataka, India. He did M.Sc. (Chemistry), PG Diploma (Pulp and Paper Technology) from Forest Research Institute Dehradoon, M.Tech. (Energy Technology) and Ph.D. from IIT Roorkee. His areas of research interests include pulp and paper industry/rayon grade pulp and fibre process engineering, process design for energy efficiency, hybrid reinforced/filled polymer composites, lignocellulosic reinforced/filled polymer composites, modification and treatment of lignocellulosic fibres and nanocomposites and nanocellulose fibres.

Dr. Anuj Kumar Verma is currently working as a Head of Quality and Product Development in Akemi Minerals, Ahmadabad. He did his B.Tech. in Chemical Engineering from NIT Warangal, M.Tech. (Exchange Program) in Pulp & Paper Engineering from IIT Roorkee and Technical University Darmstadt Germany. He did his Ph.D. from MANIT Bhopal. His areas of research interests includes specialty chemical for paper industry, artificial pulp and clay-polymeric coating.

Prof. Dr. Mohammad Jawaid is currently working as a Distinguished Professor at Chemical and Petroleum Engineering Department, United Arab Emirates University, UAE. He has been working in composite field since 2008 and has 25 years of experience in teaching, research in the field of composites. His areas of research interests include hybrid reinforced/filled polymer composites, advanced materials: graphene/nanoclay/fire retardant, lignocellulosic reinforced/filled polymer composites, modification and treatment of lignocellulosic fibres and solid wood, nanocomposites and nanocellulose fibres. Forty Ph.D. and 12 Master students graduated under his supervision in 2014–2020. He has several research grants at university/national/international level on polymer composites and its related research of around RM 6 million (USD 1.5 million).

Prof. Dr. Eldon R. Rene is currently working as an Associate Professor at IHE Delft Institute for Water Education, the Netherlands. He obtained his University Teaching Qualification (UTQ) diploma from IHE Delft and a Ph.D. in Chemical Engineering from Indian Institute of Technology Madras (India). Eldon's broad research interests are related to the development of biological treatment processes for wastewater and waste gas treatment, resource recovery, integrated urban water/waste management and the use of artificial intelligence tools for environmental monitoring and environmental process control.

Black Liquor: Chemical, Physical and Combustion Properties and Their Importance in Chemical Recovery Operations

B. P. Thapliyal and A. K. Dixit

Abstract Black liquor is an invaluable resource for paper mills, processed through various unit operations in the chemical recovery systems to recover and recycle pulping chemicals, and efficiently generate steam and electrical power. During its processing in the chemical recovery systems, black liquor undergoes four major unit operations: concentration, combustion, causticization, and calcination. These operations present significant challenges to mills, requiring careful black liquor management to ensure efficient chemical recovery and energy generation. The properties of black liquor are crucial in understanding the key characteristics that affect the operations of chemical recovery systems. These properties depend on various factors such as raw materials and cooking conditions, including chemical charge, time, temperature, and pressure. Analysis of black liquor properties helps to reveal its nature and aids plant managers and engineers in operating chemical recovery processes smoothly and efficiently. Black liquor properties can be categorized into three main subcategories: chemical, physical and rheological and combustion properties. Since the chemical recovery unit operations in mills differ in their level of operation, equipment design, automation, and controls, therefore, a better understanding of black liquor properties supports to understand their impact in efficient operation of processes. Thus, insight in the black liquor properties becomes crucial for process engineers and operators to recognize their intricate relationships with processes, allowing them to strike an optimal balance for efficient operation of chemical recovery systems.

Keywords Colloidal stability · Residual active alkali · Total suspended solids · Viscosity · Rheology · Combustion · Non process elements · Swelling volume ratio · Lime kiln

---

B. P. Thapliyal (✉)
Indian Agro and Recycled Paper Mills Association, New Delhi, India
e-mail: sg@iarpma.org

A. K. Dixit
Central Pulp and Paper Research Institute, Saharanpur, Uttar Pradesh, India

S. Sundaramurthy et al. (eds.), *Black Liquor Waste Management in the Pulp and Paper Industry*, Sustainable Materials and Technology,
https://doi.org/10.1007/978-981-96-4237-3_1

1 Introduction

Black liquor generated as the spent liquor from the washing of chemical pulp after digestion of the lignocellulosic materials, is a valuable resource that finds application as a by-product and fuel for integrated pulp and paper mills. Black liquor is processed in various unit operations of the chemical recovery systems in the mill for

- Recycling of the pulping chemicals
- Efficient generation of steam and electrical power from the black liquor, and
- Effective disposal of pollution load generated by dissolved raw material substances, both organic and inorganic materials.

In the chemical recovery process, black liquor passes through four major unit operations, concentration, combustion, causticization, and calcination. Mills face significant challenges in these unit operations to ensure efficient operation leading to improved energy and chemical recovery [1, 1–3]. Apart from the efficient operation, challenges also relate to control of air emissions, liquid effluent, solid waste discharge, etc. Hence, while the efficient operation of chemical recovery operations is important for maximum energy and chemical recovery, it is also eminent for pollution abatement from pulp making. One major contributor to the above challenges is black liquor which is a complex mixture of organic and inorganic constituents, with wide physico-chemical variations depending on the raw material, type of pulping such as kraft, soda, semi-chemical etc., chemical charge, process conditions like temperature, pressure, time etc. In view of these extensive variations and to operate the chemical recovery processes in an efficient manner, it becomes one of the most important requirements to understand the nature of a particular black liquor, by studying the physical, chemical, and combustion properties prior to its processing in different unit operations.

The physical, chemical and combustion properties of black liquors vary extensively from one pulp mill to another based on the diversities due to various reasons. It is pertinent to note that the basic principles of chemical recovery unit operations are quite similar in almost all chemical recovery operations, the only difference is in their level of operation, equipment design, automation, and controls. Therefore, without a fundamental understanding of various black liquor properties and their importance in efficient operation, it is very difficult to efficiently run any chemical recovery systems.

This chapter presents the details of various chemical, physical and rheological, and combustion or thermal properties highlighting their impact on process operations so that it is possible to evaluate and correlate them for the most efficient operation through a systematic study. The purpose of studying black liquor properties becomes more reasonable as the characteristics of black liquor vary extensively on account of the changing dynamics of a closed cycle operation in a pulp mill due to the build-up of chemicals with the addition of raw materials, makeup chemicals, process water, etc. in each cycle. Apart from the repetitive build-up of chemicals in black liquor, the properties of black liquor undergo changes with variations in raw material supply by significantly affecting heating value and overall recovery boiler thermal efficiency. Inefficient operation of the washers and recovery unit operations also impact black liquor properties due to lower black liquor solids, increased carbonate dead load, low reduction efficiency, etc. Understanding the black liquor properties, thus helps the plant technical personnel to smoothly operate the unit operations and achieve high efficiency of the chemical recovery systems.

1.1 Importance of Chemical Recovery Operations

Chemical recovery operations are the most important unit operations that help the integrated pulp and paper mill to achieve economic and environmental sustainability. From an environmental point of view, chemical recovery operations are necessary particularly to incinerate the high BOD and COD black liquor, difficult to treat in any techno-economically feasible manner. This helps to recover the inorganic chemicals as molten smelt for conversion into active pulping chemicals. The incineration of organic constituents' releases energy that can be utilized to meet the steam requirements in digesters, evaporators, and black liquor heating/soot blowing in the recovery boilers. The recovery of spent chemicals reduces the fresh active alkali requirement substantially, with the need to procure the makeup chemicals to substantiate the chemicals lost with flue gases. The steam energy generated from the heat released by the incineration of organics can significantly meet the pulp mill steam energy requirements when performed efficiently [4–6].

Figure 1 illustrates the importance of a chemical recovery system in an integrated paper mill. This emphasizes the need to improve chemical and energy recovery from the recovery operations. One of the crucial roles played in achieving efficiency and cost-effectiveness in chemical recovery operations is to understand the black liquor characteristics and subject the unit operations to meeting the most optimum operational conditions. In other words, the characterization of black liquor to understand its behaviour is the first step in attaining the best efficiency of chemical recovery operations.

2 Black Liquor Properties

Black liquor properties play an important role in understanding the nature of black liquor regarding its constituents and their effect on various operational parameters. Since the characteristics of black liquor are dynamic and change with any change in raw materials, cooking conditions like chemical charge, time, temperature, pressure, etc. Therefore, a process engineer must analyse the black liquor generated after a certain interval or on change in any process conditions. The analysis of the black liquor properties over a time span reveals the variations in its different characteristics

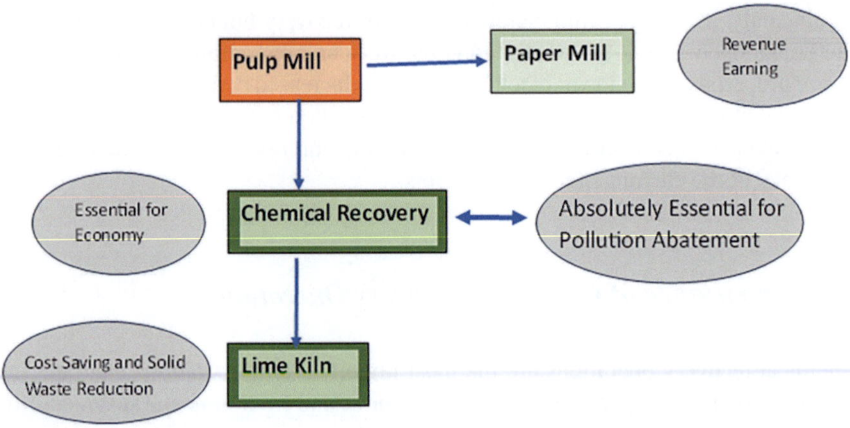

Fig. 1 Importance of chemical recovery system in a pulp and paper mill

and thereby helps the operator to operate the chemical recovery operations in a smooth and efficient manner.

The black liquor properties can be categorized into three main subcategories, such as

- Chemical properties.
- Physical and Rheological properties.
- Thermal and combustion properties

The details of various chemical, physical and rheological, and thermal and combustion properties are discussed in the next sections along with their importance, impact on recovery operations, and methods of evaluation.

2.1 Major Factors Affecting Black Liquor Characteristics

There are two important factors that affect the black liquor properties. The first is the changes in the raw material supply, resulting in changes in organic constituents and inorganic contents due to variations in chemical charge as well as the pulping conditions. The change in raw materials has a direct impact on black liquor heating value due to changes in organic constituent with change in raw materials and this in turn affects the boiler's thermal efficiency.

The other factor resulting in changes in black liquor properties is the poor operation of chemical recovery operations resulting in poor quality of white liquor, which affects the pulping chemistry and has subsequent follow-up on lower black liquor properties [7]. Therefore, an inefficient chemical recovery operation not only results in an uneconomic situation for the mill but also has a cascading effect by degrading the black liquor properties with each inefficient cycle. The uncontrolled inefficient operations of the whole system, over a period, can lead to terrible situations with a build-up of problems in almost all unit operations. The only solution to avoid such situations is to keep a constant watch on black liquor properties and address any unwanted change with immediate attention to the problem.

3 Chemical Properties

Table 1 shows the major chemical properties which are important for understanding the nature of black liquor. The chemical properties are discussed in detail in the next sections. However, it is important to understand the salient features of black liquor composition prior to going ahead with the discussion on chemical properties individually.

The above chemical properties are discussed in detail in the next sections. However, it is important to understand the salient features of black liquor composition prior to going ahead with the discussion on chemical properties individually.

The chemical properties of black liquor depend on the characteristics of the raw materials used, chemical charge and pulping conditions, etc. Once we have the overall results of the chemical analysis with us, we can review it, and find out what has happened in the recovery system. So, in other words, the understanding of black liquor chemical analysis reflects on the performance of the recovery process operations. Table 2 presents the variations in chemical properties of different types of black liquors.

The higher chemicals in black liquor composition indicate higher degradation of carbohydrates, more proportion of organic acids, and lower RAA (Residual Active Alkali) levels. Similarly higher sodium content in the black liquor indicates that the system has lower causticizing efficiency. In case black liquor has higher sodium sulphite, it indicates lower reduction efficiency and the addition of excess makeup chemicals. It is quite evident that a thorough investigation of black liquor chemical

Table 1 List of important black liquor chemical properties

• Total solids
• Total suspended solids
• pH
• Residual active alkali
• Total alkali
• Ratio of organic to inorganic compounds
• Elemental composition of black liquor solids
• Colloidal stability
• Non-process elements and their impact on recovery operations

Table 2 Properties of different black liquors

Parameters	Unit	Bagasse	Wheat straw	Rice straw	Wood
pH		12.5	12.3	11.2	12.5–13
Total Solid	% *w/w*	13–16	13–16	16–18	16–19
Inorganics as NaOH	% *w/w*	33–35	34–36	35–36	32–37
Organics	% *w/w*	65–67	64–66	65–64	63–68
Chlorides as NaCl	% *w/w*	0.7–0.9	1.3–1.7	1–1.2	4–5
Silica as SiO_2% *w/w*	% *w/w*	1–1.5	2.5–3.0	0.05–0.1	0.05–0.15
Residual active alkali	g/l as Na_2O	5–7	5–7	5–6	4–7
Swelling volume ratio	ml/g	13–15	9–13	20–25	25–30
Calorific value as GCV	cal/g	2900–3000	2800–2900	2700–2800	3000- 3200
Carbon as C	% *w/w*	34–35	32–33	32–34	30–32
Hydrogen as H	% *w/w*	4.1	2.9	3.2	3.6
Nitrogen as N	% *w/w*	0.3	0.4	0.4	0.01
Sulphur as S	% *w/w*	0.5	0.5	2	2.5–3

Source CPPRI database

analysis can reveal a lot of information's about what is wrong in the process operations and how it can be set right [8, 9].

3.1 Total Solids

The concentration of solids in weak black liquor is an important parameter that governs the steam economy in evaporators. It is in fact one parameter that denotes a theoretical indication of the total steam requirement for raising the solids concentration from a lower level to a higher level in the evaporators. The theoretical calculation can be compared with the actual steam requirement in the evaporators to find out the efficiency of evaporation. In case it is lower, than appropriate corrective action can be initiated to increase the evaporator efficiency.

Additionally, the efficiency of the washer can also be increased based on the total solids analysis. An efficient washing operation generally results in high solids of weak black liquor, and it can be reviewed during the process of washing.

In actual sense, for every 1% increase in weak black liquor dry solids, because of washing efficiency improvement, the mill can save 40–45 tons of steam during evaporation in a 100 TPD pulp mill. Table 3 shows the approximate volume of water evaporated at various black liquor concentrations.

Table 3 Approximate volumes of water evaporated at various black liquor concentrations

Type of black liquor	Conc. of W.B.L. (% *w/w*)	Evaporated water, t/t pulp (at 50% Solids)
Wood	14–17	6.6–5.9
Bagasse	7–10	8.0–10.4

Methods of Total Solids Determination

Total solids content of weak black liquor virtually decides the steam requirement in evaporators. This is an important parameter used to calibrate routine control parameters in chemical recovery. This method measures the solids remaining after removal of water and other non-aqueous volatile materials normally lost in the evaporation section.

Procedure [10]

25–30 g of sand is placed in a glass weighing bottle with a cap and kept in the oven at 105 °C to a constant weight. It is cooled in a desiccator and weighed to the nearest 0.5 mg. By means of a pipette, enough black liquor specimen is transferred to the weighing bottle ensuring nearly 1 g of dry solids. The cap is put on the bottle and weighed to the nearest 0.5 mg. The weighing bottle is kept in the oven without the cap for a minimum of 6 h. The weighing bottle is then removed and cooled in a desiccator and weighed with the cap on. This procedure is repeated at one hour interval until the weight loss is less than 0.1% per hour. The calculation of the total solids is as follows:

$$\text{Total solids, }\%\ w/w = \frac{\text{Weight of dried solids} \times 100}{\text{Weight of specimen}}$$

3.2 *Total Suspended Solids*

Suspended solids determination in black liquor is also one of the parameters to process it in a smooth way during the chemical recovery operation. The suspended organic particles in the black liquor are the main cause for the increase in viscosity during concentration in evaporators. The suspended particles enhance scale formation during evaporation in the evaporators as they act as initiation points for precipitation when the black liquor is concentrated.

Higher viscosity and scale formation limit the final concentration in the evaporator outlets for firing in the recovery boiler and thus it results in lower thermal efficiency of the recovery boiler. As a good practice, a 20-ppm level of suspended solids is good enough for the smooth operation of black liquor evaporators. The 20-ppm level of suspended solids can be achieved in the normal course by passing the black liquor from filters or fine screens to make the black liquor free from suspended materials before taking it to the chemical recovery operations.

3.3 pH of Black Liquor

The pH of black liquor is very important for colloidal stability of the black liquor, as at a lower pH level lignin starts to precipitate from the black liquor. It is observed that alkali lignin in black liquor is colloidally stable above pH 12.0, and it starts to precipitate when pH is reduced to a level of 9.0–10.0. It is reported that by lowering the pH, the polymerization of saccharinic acids leads to an increase in black liquor viscosity. Therefore, for maintaining the colloidal stability, a pH value above 12.0 is required [11].

3.4 Residual Active Alkali

The residual active alkali measures the unspent NaOH and Na_2S after the completion of the pulping operation. This is an important parameter and must be maintained to avoid the precipitation of lignin from the black liquor. At least 6 g/l residual alkali should be maintained by controlling the pulping conditions and chemical charge. The level of free NaOH and Na_2S in the liquor is influenced by the type of raw material and pulping conditions therefore, residual active alkali that determines their presence, is considered important from the viewpoint of colloidal stability [12]. Normally the measurement of RAA is made by titrimetric analysis, but it is now determined by automatic potentiometric titrations. Milonava and Dorris [13] have conducted studies on determination of RAA using model solutions and various black liquors by potentiometric titrations. Various other researchers have also contributed to RAA determination [14, 15].

Higher alkali levels increase the swelling volume ratio (SVR) of the black liquor in the initial stage, but at very high residual alkali levels swelling volume index suddenly drops during the combustion process [16, 17]. The effect of residual active alkali is also prominent in viscosity control as on the viscosity decreases with its increase.

In previous works, different studies have reported the effect of residual active alkali on the viscosity in different types of wood-based black liquors [17, 18]. Marimuthu et al. [18] have reported the behaviour of bagasse and wood black liquors and evaluated the effect of residual active alkali on viscosity of both types of black liquors. They reported that the viscosity of bagasse black liquors is higher at all RAA levels (3.0, 5.0, and 7.0 g/l) when compared with wood black liquors. The SVR values for bagasse black liquor reported to be lower than wood black liquors.

In a recent study by Silva et al. [19], black liquor samples were produced from *Ochroma pyramidale* wood over a range of 10–24% effective chemical charge with an interval of 2%. In these black liquor samples, the RAA and pH were related positively to the increase of the alkali charge during pulping. However, the increase in alkali charge decreased the pulp yield, kappa number and rejects. The hexanuronic acid concentration was found to increase on the other hand.

Estimation of residual active alkali in black liquor

It is the measure of free NaOH and Na_2S contents remaining in the black liquor after pulping. Certain level of residual active alkali is essential to keep the black liquor colloidally stable. Liquors with low levels of RAA may precipitate at low solid concentration during evaporation, which eventually will increase the black liquor viscosity.

Procedure [20]

25 ml of black liquor is pipetted into a 500 ml volumetric flask containing about 100 ml of water. 20 ml of 0.5 *N* $BaCl_2$ is added, and the contents stirred and allowed to settle until partially clear. The contents of the flask are diluted to the mark, mixed well and allowed to settle until about 200 ml of clear liquid stands above the precipitate. 100 ml of the supernatant liquid is pipetted into a 250 ml beaker equipped with a magnetic stirrer. 5 ml of 40% neutral formaldehyde is added and titrated immediately with 0.1*N* HCl to a pH of about 3.5 following the course of titration potentiometrically on the pH range of a pH meter using standard calomel-general purpose glass electrode pair. The titrant is added in increments of fixed volumes, the pH value is recorded after each addition and the data is plotted on the linear coordinates with Δ pH/ΔV as the ordinate and milliequivalents of HCl as the abscissa. The plotted points are connected using a smooth curve. Milliequivalents of acid corresponding to the end point by projecting the inflection point near pH 7.0 to the abscissa are obtained.

Calculations

$$\text{R.A.A g/l as Na}_2\text{O} = \frac{\text{milliequivalent of acid} \times 0.1 \times 31}{\text{ml of black liquor taken (5 ml)}}$$

To determine the residual active alkali in soda black liquor, addition of formaldehyde is not required and results are reported as g/l NaOH. In this case, the RAA is calculated as follows:

$$\text{R.A.A g/l as NaOH} = \frac{\text{milliequivalent of acid} \times 0.1 \times 40}{\text{ml of black liquor taken (5 ml)}}$$

3.5 *Total Alkali*

It is a term used to show the collective measure of all the inorganic and organic sodium salts present in the liquor. It includes NaOH, Na_2CO_3, organically bound sodium, and silicates. Sodium carbonate content is important as it has been observed that when present in excess quantity, it forms double salts which havc inverted solubility and deposits in the form of scales. The total alkali measurement is important for calculating the efficiency of the chemical recovery cycle.

3.6 Ratio of Organic to Inorganic Compounds

The presence of organic and inorganic constituents in black liquor affects the properties and behavior of black liquor to a very large extent. The ratio of organic to inorganic constituents is influenced by chemical charge and resulting pulp yield. Organic to inorganic ratio determines properties like specific gravity, surface tension, specific heat, and viscosity. Higher organic matter in black liquor results in higher heating values, whereas the high Inorganic content (more than 35% as NaOH) leads to decreased critical solids which cause scaling in the evaporator.

In the kraft process, the chemical charge during pulping containing hydroxide and sulfide mineralizes the carbon and oxygen from the organic portion of the black liquor and converts it into carbonate and sulfate during the cooling and ashing operation. As a result, the inorganic contents of black liquor increase and it causes problems in estimating the total black liquor production. Andrew et al. [21] have used a common factor to correct the lab test results to provide a more realistic estimate of the initial inorganic used and the organics removed in the digester to estimate black liquor production.

In a study, Su et al. [22] reported that in mixed black liquors, the ratio of the raw materials used for pulping can also affect the combustion properties significantly mainly due to changes in organic and inorganic ratios and SVR of the black liquor.

Estimation of organic to inorganic ratio

Estimation of sulfated ash in black liquor represents the organic to inorganic ratio of the black liquor. It also indicates the total recoverable sodium present in the black liquor, which is estimated in the form of sodium sulfate (Na_2SO_4).

Procedure [23]

In a weighed and tared platinum crucible, 2–3 g of exactly weighed (A) residue/ black liquor solids are taken. Starting with cold Muffle furnace, the residue is heated at 600° C until the organic material is mostly, but not destroyed. The residue is then cooled and moistened with water. To this, a drop of methyl orange is added followed by concentrated H_2SO_4 drop-wise till the indicator color turns pink. The contents in the crucible are first dried on water/steam bath and then in Muffle furnace at 800 °C till the residue turns into white–colourless material. The residue is then cooled in a desiccator and weighed. After weighing, the residue is again moistened and a drop of methyl orange is added to see whether residual H_2SO_4 is left. If pink color is developed (indicating the presence of residual H_2SO_4), the above procedure is repeated until the acid is evaporated completely. The residue is arbitrarily assumed to be pure Na_2SO_4 and is weighed again (B).

The percent of black liquor residue converted to Na_2SO_4 is expressed as NaOH and it gives an estimation of the inorganic matter present in the black liquor. The remaining part (i.e. 100-inorganics) is the percent organics present in the black liquor, as shown below.

$$\text{Weight of black liquor solids taken : A}$$

$$\text{Weight of } Na_2SO_4 : B$$

$$\text{Inorganics i.e. sulfated ash as NaOH (as \% of total solids)}$$

$$\text{Inorganics, \% as NaOH} = \frac{B \times 40 \times 100}{A \times 71} = \frac{B \times 56.4}{A}$$

$$\text{Organic matter, } \%w/w = 100 - \text{Inorganic matter (By difference)}$$

3.7 *Elemental Composition of Black Liquor*

The elemental composition includes the estimation of C, H, N, O, S, and Na. Elemental composition is considered important in terms of heating value and subsequent combustion characteristics of spent liquor.

Elemental composition, particularly C, H, N, and O is very important in the design and development of recovery boilers [20]. Estimation of other elements like Fe, Al, Cr, Ca, and Mg (NPE) is also important as most of these elements enter through process water and/or by corrosion of equipment and they deposit in the form of scales on evaporator tubes thereby reducing the heat transfer efficiency.

3.8 *Colloidal Stability*

Spent liquor is a colloidal solution containing two major colloidal systems namely molecular colloids formed by the alkali lignin and association colloids formed by the fatty acids. The stability of these colloidal systems is affected by pH of the black liquor and in the low pH i.e., below 10.0, the colloidal particles tend to aggregate due to charge neutralization leading to precipitation.

Apart from the pH, colloidal stability of black liquor is influenced by other factors also, like the high proportion of higher molecular weight lignin macromolecules and the presence of organic acid salts in significant proportion. Colloidal stability is considered important from the viewpoint of efficient and smooth operation of evaporator units to achieve higher solids for firing the black liquor in the recovery boiler.

3.9 Non-process Elements and Their Impact on the Recovery Operation

Non process elements (NPEs) do not play any role in pulping and chemical recovery operations. In fact, they create serious problems in smooth and efficient operation of the chemical recovery systems. The major NPEs in chemical recovery systems are chloride, silica, and potassium. The major sources of these NPEs are raw materials and process water. Tables 4 and 5 show the NPE levels present in various raw materials.

Operation of various unit operations like the multiple effect evaporators, recovery boiler, causticization, and lime reburning in lime kiln are significantly influenced by the presence of NPEs in the black liquor and green liquor.

Table 4 NPEs in various raw materials

Particulars	Bagasse	Wheat straw	Rice straw	Sabai grass	Hardwood
Ash, % *w/w*	2.3	9.5	18	8.5	0.3
Silica, % *w/w* as SiO_2	1.8	5.8	11	6.5	0.2
Potassium, % *w/w* as K	0.2	1.8	2.0	1.5	Tr
Chlorides, % *w/w* as Cl	0.02	0.88	0.94	0.3	0.05

Source CPPRI database

Table 5 Level of NPEs in different hardwoods

Name of NPE (%)	Casurina	Cashew	Popular	Subabul	Eucalyptus
Silica	0.06	0.03	0.03	0.02	0.08
Calcium	0.1	0.2	0.02	0.2	0.07
Potassium	0.11	0.26	0.2	0.25	0.27
Chlorides	0.03	0.03	0.02	0.04	0.02

Source CPPRI database

3.9.1 Impact of Chlorides and Potassium on Chemical Recovery Systems

The two major non-process elements found to have the most significant influence on recovery boiler operation are potassium and chlorine present as chlorides. It is apparent from the studies that they are volatile at relatively lower temperatures and have an adverse effect on ash melting temperatures [24].

A high potassium level in the liquor aggravates corrosion in the super heater area and increases fouling in the recovery boiler, particularly when it is also accompanied by a high chloride level [24]. Chlorides have been observed to seriously impact recovery boiler fouling and plugging. Chlorides also have been related to some of the corrosion problems around the recovery boiler and other unit operations of chemical recovery systems.

3.9.2 Impact of Silica and Other NPEs

Silica is also one of the NPEs which significantly affects almost all operations of pulp mills. Silica is associated with scaling problems as it acts as one of the nucleating agents in the presence of other NPEs to initiate scaling problems on evaporators, filters, wires, screens, furnaces, etc. It results in higher smelt melting temperature and creates problems of freezing near the spout leading to problems in restriction of smelt flow. Silica in the white liquor leads to poor quality due to the low settling tendency of the sodium silicates in the settling tanks of the causticization sections. Silica present in the lime sludge results in poor quality re-burnt lime quality mainly leading to over-burnt lime. It also results in various other operational problems in lime mud re-burning in the lime kiln [25–27].

Estimation of silica in black liquor

The essence of the method is the destruction and removal of all organic matter and the oxidation of all inorganic material, including sulfur, to its highest valence by digesting it with HNO_3 and $HClO_4$. On digestion, a clear solution containing all the inorganic constituents of the original liquor having silica in precipitated form is obtained.

Procedure [28]

10 ml black liquor is pipetted into a 250 ml beaker, allowing the pipette to drain for its full calibration period. To this black liquor, 5 ml of conc. HNO_3 and 5 ml of conc. (70%) $HClO_4$ is added in that order. The flask is placed on a water/steam bath inside a hood. When the oxidation reaction has subsided, as indicated by the absence of red-brown nitrogen dioxide fumes, the flask is removed from the water bath and heated over a hot plate with low to moderate heat until the oxidation is complete, as evidenced by a colorless solution and dense white perchloric acid fumes. The solution is allowed to cool, 5 ml concentrated HCl is added, and heated again till a slurry is formed. The solution is cooled to room temperature and 50 ml of warm

distilled water is added to dissolve all the inorganics but silica. The solution is filtered through Whatman filter paper no. 41 and the residue is washed with 1:99 HCl followed by hot water (2–3 times) washing. The filtrate and washings are kept reserved for determination of iron and aluminum (mixed oxides i.e. R_2O_3).

The filter paper is placed in a previously ignited and weighed platinum crucible. The filter paper is dried and smoke off at low heat in the furnace. When the paper has been charred and smoking has almost ceased, the temperature of the furnace is increased to 600 °C and heated at this temperature for 1.5 h. The crucible is taken out and placed in a desiccator and weighed, reheated, and weighed again till obtaining a constant weight (W1).

The contents of the crucible are moistened with about 2 ml of water, adding it with a pipette. 4–5 drops of conc. H_2SO_4 are added and the crucible is placed in a hood over a hot plate to fume off the H_2SO_4, then about 5 ml of concentrated Hydrofluoric acid (HF) measured from a polythylene graduated cylinder is added and heated gently to evaporate off the HF without boiling the liquid. When the contents of the crucible are evaporated to near dryness, the crucible is kept in the furnace and gradually the temperature is increased to 800 °C. Constant weight of the crucible is achieved as stated earlier (W2). The loss in weight represents the grams of SiO_2 in the 10 ml specimen.

Calculation

$$\%\text{ of silica as } SiO_2 = \frac{SiO_2(W1 - W2) \times 100}{\text{g of black liquor solids in 10 ml sample taken}}.$$

Estimation of mixed oxides (R_2O_3) in the black liquor

The mixed oxides or R_2O_3 refers to the estimation of aluminum and iron in combined form as their oxides. Iron (Fe) and aluminum (Al) are the non-process elements present in black liquor. Estimation of these elements is important as they deposit in the form of scales on evaporator tubes thereby reducing the heat transfer efficiency.

Procedure [27]

The filtrate, after the estimation of the silica content in black liquor is collected and heated to boiling. Few drops of methyl red indicator are added followed by the addition of conc. NH_3 solution drop-wise with constant stirring until the solution become alkaline. One drop of ammonia is added in excess and the solution is heated for a few minutes till the precipitate flocks. The contents are filtered through a Whatman filter paper no. 42 and the precipitate is washed with 1% NH_4Cl.

The filtrate and washings are collected in a 250 ml volumetric flask to further proceed for sulfur estimation. The precipitate is ignited in a previously ignited and weighed (Weight A) platinum crucible at 800 °C in a Muffle furnace to a constant weight (B).

Calculation

$$R_2O_3, \%\ w/w = \frac{B - A \times 100}{\text{g of black liquor dry solids taken initially}}.$$

Estimation of sulfur in the black liquor

Elemental analysis provides the main chemical components in the black liquor. The elemental composition includes estimation of C, H, N, O, S and Na. Elemental composition is considered important in recovery boiler material and energy balance calculations. This method provides the procedure for the estimation of total sulfur in black liquor gravimetrically.

Procedure [27]

The filtrate obtained after precipitating mixed oxides or R_2O_3 is collected in a volumetric flask and made up to 250 ml volume. To this solution concentrated HCl is added drop-wise till the solution turns pink with the addition of methyl orange indicator. 1 ml of the acid is added in excess. The contents are heated to boiling conditions and 15 ml of hot 20% $BaCl_2$ solution is added into it with constant stirring. Keeping the stirring on for few minutes and the solution is allowed to stand for an hour. Filtered through Whatman filter paper no. 42 and the precipitate is washed thoroughly with warm water. The absence of chloride ions is tested with silver nitrate. The precipitate is ignited in previously ignited and weighed (Weight A) platinum crucible at 800 °C and the residue is finally weighed as barium sulfate (B).

Calculation

$$\text{Sulfur}, \%\ w/w = \frac{\text{g of residue as } BaSO_4(B - A) \times 32.1 \times 100}{233.4 \times \text{g of black liquor dry solids taken initially}}.$$

The NPEs, other than potassium, chlorides, and silica, also have adverse impacts on chemical recovery operations. The impacts of all NPEs which can affect the various operations of chemical recovery systems are summarized and listed in Table 6.

4 Physical and Rheological Properties

The Table 7 shows the major physical and rheological properties of black liquor which influence its characteristics and therefore impact the efficiency of chemical recovery operations.

4.1 Viscosity of Black Liquor

Viscosity is a very important property for black liquor handling (pressure drop), heat transfer, and mixing operations. It is necessary to keep black liquor in colloidal

Table 6 Impact of NPEs on chemical recovery system

NPEs	Impact on chemical recovery system
Silica (Si)	- Scaling on evaporators, filters, wires, screens, furnace etc. - Higher smelt melting temperature, more problem of freezing near the spout - Poor white liquor quality - Poor re-burnt lime quality (over burnt lime) - Operational problems in lime mud re-burning
Chlorine (Cl)	- Increasing scaling and corrosion rate - Adversely affects ash fouling characteristics, lower sticky temperature on heat transfer surfaces
Calcium (Ca)	- Increases water soluble scales on evaporator tubes - Poor white liquor quality
Iron (Fe)	- Formation of corrosion and scaling products like Fe_2O_3,FeS and $FeSO_4$ on vapor side evaporator tubes - Presence of Fe_2O_3 dust on economizer tubes accelerates SO_3 formation, leading to NaHSO4 formation
Aluminum (Al)	- Scale formation—along with Na and Si, scaling is increased on evaporator tubes
Phosphorous (P)	- Affects re-burnt lime quality - Operational problems due to ring formation in lime kiln
Potassium (K)	- Poor white liquor quality - Adversely affects ash fouling characteristics, lower sticky temperature on heat transfer surfaces

Table 7 Physical and rheological properties of black liquor

Viscosity of black liquor
Boiling point elevation
Fouling tendency
Surface tension
Colloidal stability at higher concentration

stability and free flowing in concentration and combustion stages of the Chemical recovery process. Black liquor viscosity depends on the nature and amount of polymeric materials (lignin, extractives, and hemicellulose), the amount of alkali used in cooking, and pH. Viscosity is an important property for black liquor pumping, heat transfer, and mixing operations. To maintain the black liquor in flowable condition without any problem, it is necessary to keep black liquor in colloidal and free-flowing conditions [29].

The weak black liquor behaves as Newtonian fluid at lower solids levels and as the concentration of black liquor increases during the evaporation in evaporators, it starts to behave as non-Newtonian fluid. Strong black liquor behaves as a non-Newtonian liquid in characteristics and its viscosity is found to be a function of concentration, temperature, and pH.

Zaman and Fricke [30] reported correlations for the viscosity of kraft black liquors at low solids concentrations. In another study [31] they studied the effects of solids content, degree of delignification, and black liquor composition on the viscosity of kraft black liquors at low solids concentrations. Further, Zaman et al [32], also studied the linear viscoelastic functions of several softwood kraft black liquors at higher solids concentrations, from 65% to 81% at temperatures from 40 to 85 °C. They reported that at high solids and lower temperatures, black liquors behave like uncross-linked polymers that are dependent upon the solid's composition varying from liquor to liquor. Several researchers have also significantly contributed to the flow behavior studies of black liquors [33–35].

Table 8 shows the changes in the viscosity of different black liquors as a function of solids concentration at a fixed temperature of 100 °C.

Residual active alkali also plays an important role in the viscosity of the black liquor due to its impact on colloidal stability. According to the relationship between viscosity and total solids at different RAA levels for different types of black liquors, the non-wood black liquors have higher viscosity when compared with bamboo and tropical pine black liquors.

An increase in viscosity by a factor of 10, can decrease evaporator capacity by 0.5 times. Similarly, a decrease in viscosity by a factor of 10 can increase evaporator capacity by 1.5–2.0 times. Black liquor viscosity increases significantly at higher solids concentration and lower temperatures, therefore, ratio of viscosity of black liquor at two solids concentration at a particular temperature relates with the heat transfer efficiency in the evaporator, showing a decrease in relative capacity with increase of viscosity ratio, depending on factors like heat transfer efficiency, flow dynamics and the physical properties of black liquor.

Pyrolysis and combustion of black liquor in recovery boiler is dependent on viscosity and temperature of black liquor. During pyrolysis and combustion, the size of black liquor droplets plays a critical role due to resistance to atomization at higher viscosity leading to formation of larger droplets during black liquor spraying, resulting in incomplete combustion, increased char bed height, and fouling of furnace walls. While optimizing droplet size through proper nozzle design and operating

Table 8 Viscosity of different black liquors

Raw material	Viscosity, mPa.s at 100 °C at Different Total Solids *%w/w*				
	45	50	55	60	65
Pine	9.1	18	40	160	215
Eucalyptus	2.3	83	380	490	640
Eta Reeds	41	100	395	470	615
Bamboo	2	11	31	125	398
Wheat Straw	40	89	299	450	1000
Bagasse	125	161	912	1100–1250	3100–3300

Source CPPRI database

conditions by lowering viscosity and increasing temperature of firing liquor is essential for maximizing combustion efficiency, the droplet size must be balanced to avoid excessive carryover or other operational issues caused by smaller droplets formation. It indirectly shows the effect of viscosity as it is dependent on the solids concentration and temperature of black liquor.

Measurement of viscosity of black liquor

Black liquor is a colloidal system and the viscosity depends largely upon the colloidal stability of the lignin macromolecules present. Viscosity of black liquor is an important aspect in its processing for chemical recovery. Higher pH and presence of free alkali ensures that lignin remain in colloidal form and low viscosity of the black liquor.

Procedure

Viscosity of the black liquor is measured by using HAAKE Viscometer at 90 °C and at various total solid concentrations. Temperature control is accomplished by using hot fluid circulators, such as the JULABO circulators.

Preparation of black liquor at different total solid concentrations is also an important activity, before the viscosity determination, and it requires a precise control. The method of obtaining black liquor at different solids for viscosity estimation is explained below.

Black liquor from laboratory cooks has 12–20% solid concentrations. The liquor needs to be concentrated in a rotary vacuum flash evaporator (Fig. 2) under vacuum. The rotary vacuum flask is heated in an electrically heated oil bath maintained at a temperature of 96–99 ^{0}C for preparation of black liquor solids concentrations from 35 to 70%. A 500 g sample of weak black liquor is placed in the flask and heating continued until the desired amount of water is removed leaving black liquor having targeted solid concentration. Sample is drawn at this point and remaining black liquor is further concentrated in the same manner for higher solid concentration. The process is continued to achieve all other desired solids concentration levels.

Measurement of viscosity

Analog or digital viscometers are used to determine the viscosity of the black liquor at different solids levels. HAAKE viscometer (Fig. 3) is one of the equipment which can be used to measure the viscosity of black liquor at different concentration.

For the determination of viscosity by a viscometer, 8–10 ml of sample is taken for measurement as per the suppliers manuals of the equipment. Temperature is maintained at 90 °C. The samples are collected after viscosity measurement from the adapter of the viscometer. The same are diluted and total solids determined as per standard method mentioned earlier.

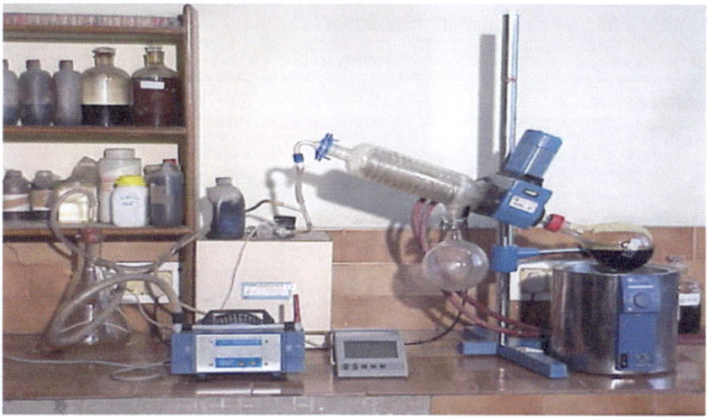

Fig. 2 Photograph of rotary flash evaporator

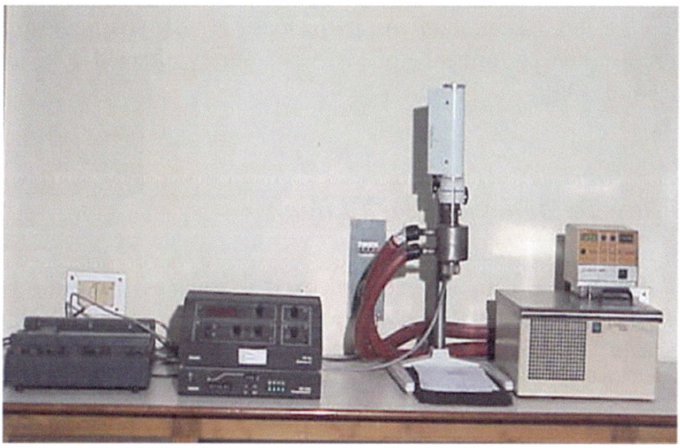

Fig. 3 Photograph of HAAKE viscometer

4.2 Boiling Point Elevation (BPE)

Black liquor is a complex colloidal mixture of organic and inorganic constituents, and all these constituents have direct and indirect effects on the thermal behavior of the black liquor. One of the important effects is the boiling point of the black liquor. It is observed that as the molar concentrations of inorganic and organic constituents increase the boiling point also increases and is termed as BPE with increasing solids concentration of black liquor.

An analysis of BPE has shown that it increases linearly with an increase in the molar sum of Na and K in black liquor. It is also dependent on Pulping variables (Chemical charge, Pulp yield, Liquor causticity, Liquor sulfidity, Chloride content, Sulfate reduction, etc.).

Table 9 Foaming tendency of different black liquors

Black liquor	pH	Foam index	Foam coefficient (kf)
Bagasse (depithed)	11.0	10	6
Bagasse (whole)	11.0	14	12
Rice straw	10.0	15	14
-do-	11.0	18	20
Wheat straw	10.0	15	15
Wood	11.1	7	18

Source CPPRI database

In MEE where the concentration of black liquor is raised from lower to higher solids, the performance of each effect of the evaporator also changes due to BPE. Therefore, BPE has become an important property and plays a vital role in the design and operation of black liquor evaporators. Studies have shown that BPE increases with an increase in concentration and its increase is very sharp above 50% total solids. Therefore, the performance of MEE is severely affected at concentrations above 50% solids [36].

4.3 Fouling Tendency of Black Liquor

Foaming is one of the common phenomena that happens when black liquor is subjected to evaporation. It takes place mainly due to air entrainment and is also witnessed during pulp washing. Foaming leads to increased corrosion and scale formation. It increases due to the presence of organic acids, gums, polymeric compounds, resin acids, etc. in the black liquor. Various defoamers like sulphonated castor oil, Turpentine, Kerosene, and Cotton seed oil are used to control foam. Table 9 shows the foaming tendency of various black liquors measured as the foaming coefficient.

4.4 Surface Tension (ST)

Surface tension is important in determining the droplet size at the time of atomization of black liquor, combustion, and nucleate boiling. Viscosity and surface tension influence spray size distribution, droplet combustion and significantly affects the performance of the recovery boiler. It decreases with an increase in concentration and temperature of black liquor. Further, at a fixed temperature, surface tension is observed to drop with increasing solids content. Basically, the organic constituents in the black liquor strongly influence surface tension.

5 Thermal and Combustion Properties

Table 10 shows the thermal and combustion properties of black liquor which play an important role during incineration of black liquor in the recovery boiler.Combustion of black liquor is a complex step and involves stages like drying, pyrolysis and char burning. The thermal efficiencies as well as recovery efficiencies are determined by the way in which each step is accomplished under the conditions prevailing in the recovery furnace. Black liquors generated from different raw materials and pulping conditions exhibit differently in the recovery furnace.

Combustion in general is influenced by two factors- those arising from poor black liquor quality and those attributed by incorrect operating variables. The later factor is largely determined by the former factor. Thermal properties i.e. swelling characteristics, calorific value of black liquor and their rate of mass loss determine their trajectories in a recovery boiler. Therefore, it is important to study the thermal properties of black liquor in order to see the effect of desilication on thermal properties of black liquor. Swelling volume ratio, calorific value and thermogravimetric analysis of the black liquor are discussed in this section.

5.1 Calorific Value (CV)

The calorific value is strongly influenced by black liquor composition (elemental analysis). It indicates the amount of heat released during combustion in the bomb calorie meter. Organic contents contribute to calorific value while Inorganic contents act as diluents. Pulping parameters (AA, Yield etc.) change the CV of BL solids. The calorific value would depend on the proportion of organic constituents and the carbon content in particular which accounts both for fixed carbon and volatile carbon.

5.1.1 Measurement of calorific value of black liquor

Calorific value of black liquor solids is an important data for the energy balance of the recovery boiler and for determining the amount of high-pressure steam that can be

Table 10 Thermal and combustion properties of black liquor

Calorific value (CV)
Swelling volume ratio (SVR)
Thermo-gravimetric analysis (TGA)
Temperature of ignition (TIG)
Integral procedural decomposition temperature (IPDT)

generated in chemical recovery boiler. The calorific value depends on the properties of organic constituent and the carbon content in particular.

Procedure [37]

Black liquor is dried an oven at 105 °C over night. A little less than 1 g of sample is pressed into pellets by means of a pellet press. This is done to prevent scattering of material during the combustion, with consequent incompleteness of reaction.

The pellet is weighed and placed in the sample pan. The fuse wire of measuring length about 7 inch of known combustion per unit length is attached to the two terminals and adjusted to give firm contact with the pellet. The cover is carefully assembled with the bomb and tightened. The bomb is then connected to the oxygen tank and oxygen is admitted slowly until the pressure of 15 psi is attained. 2030 g water (the temperature of which is adjusted so as to be at least several degrees below the upper limit of the thermometer range and close to room temperature) is put in the calorimeter can, and placed with the adiabatic jacket. The ignition leads are connected and the bomb is immersed in the water.

The cover of adiabatic jacket is set in place and the thermometer lowered into position. The stirrer is started and the jacket temperature is then adjusted to within 0.01 °C of that of water in the can. The thermometer in the can is read for a few minutes to be sure that equilibrium has been attained. This temperature is recorded as the initial temperature (T_1). After the successful ignition, the temperature of calorimeter rises quickly. The jacket temperature is kept close, as much as possible, to that of the can. The final steady temperature of the can is then recorded as T_2. The bomb and calorimeter are carefully cleaned and dried after each experiment.

Calculations

$$\text{Calorific value (cals/g)} = \frac{\text{Water equivalent } \times (T_2 - T_1) \text{ of black liquor}}{\text{Weight of oven dry sample in g}}.$$

5.2 *Swelling Volume Ratio (SVR)*

SVR estimation is a semi-qualitative method to assess the burning characteristics of black liquor solids. It reveals the height of the char bed in the combustion zone of the recovery boiler. SVR is defined as the volume of black liquor solids when roasted at 400 °C for 2 min as ml/g.

High SVR indicates a porous ignited bed with free air passage and complete incineration. Low SVR is attributed to higher inorganic load, soap content, black liquor aging, and higher extractives.

Estimation of swelling volume ratio (SVR) of the black liquor

Swelling volume ratio (SVR) measurement is an empirical and semi-quantitative method to assess the burning characteristics of the black liquor. More the swollen volume ratio better the rate of thermal decomposition. For a good burning spent liquor, freshly softened mass must have the characteristics of viscous gaseous vapor that can push the film and finally escape leaving numerous voids in swollen mass. A good burning black liquor forms a fine homogenous film on evaporation, while poor burning liquors leads to form a hard heterogeneous film of low plasticity.

Procedure [38]

A known volume of sample containing approximately 1 g solids of black liquor is taken and placed in a tared 100 ml capacity porcelain evaporating dish and weighed. The sample is slowly evaporated to dryness by keeping overnight in an oven at 105 °C. The dish is then placed in a Muffle furnace heated to a temperature of 400° ± 10 °C, exactly for 2 min. The swollen mass is examined carefully after cooling and the porous voids are sealed with a thin layer of wax. The dish is then filled with small plastic beads (0.5 mm diameter). The surface is levelled off by removing the excess beads slowly, moving a glass rod across the edge of the dish. The volume of the beads above the surface of the swollen black liquor is measured by placing the beads in a measuring cylinder (A). The volume of the empty dish after scrapping off the swollen mass is also measured using the same beads (B).

Calculation

$$\text{Swollen volume} = \text{Vol. of empty dish (B)} - \text{Vol. above the swollen mass (A)}$$

$$\text{Swelling volume ratio (SVR ml/g)} = \frac{\text{Swollen volume in ml}}{\text{Total solids present in the sample taken (g)}}.$$

5.3 *Combustion Properties - Evaluation of Thermal Behavior Using Thermogravimetry.*

In thermogravimetry, the change in sample mass is recorded as a function of temperature. It provides important information about the combustion behavior of black liquor samples over a wide range of temperatures. The analysis of thermogravimetric curves reveals drying, pyrolysis, combustion, and smelt formation phases taking place in the furnace. This gives a fairly good idea about the thermal behavior of different types of black liquor samples under controlled heating conditions.

A literature search from some previous studies shows the pyrolysis of kraft black liquor from wood which is determined after analysis of the composition gases as a function of time and temperature [39, 40]. Thermogravimetric studies have also been carried out on kraft black liquors to evaluate the behavior of different organic components during devolatilization [41]. Different liquors behave very differently in the same recovery boiler under the same operation conditions; therefore, each type of black liquor requires a specific study to evaluate its pyrolysis behavior to determine the optimal conditions for incineration or pyrolysis [42].

Recent interest is in the rapid pyrolysis of black liquor from wood at high temperatures and high pressures using pressurized reactors and many researchers are working on thermogravimetric methods for these evaluations. Gea et al. [43] carried out thermal devolatilization studies of alkaline black liquor from straw using thermogravimetric methods in different inert atmospheres. They quantified the effect of several operating conditions on the thermal devolatilization of the straw black liquors.

Thermogravimetric-Fourier transform infrared spectrometric analysis and TGA–DSC have also been used for gasification studies of black liquors [44–46]. IPDT i.e., integral procedural decomposition temperature, is the temperature at which about half of the mass is burnt. Higher IPDT values indicate slow decomposition of the matter [47]. TIG i.e., ignition temperature is the temperature at which the char burning starts. When the liquor solids are heated at sufficiently high heating rates in a temperature-programmed furnace there is a sudden change in the slope of the TG curve and a simultaneous release of heat is observed on the DTA curve. The point corresponding to the DTA signal is taken as ignition temperature [48] (Table 11).

Thermo-gravimetric analyses of black liquor

The TGA and DTA curves of black liquors can be evaluated on a variety of instruments available in the market. We have used the Stanton Redcroft STA 781 (Fig. 4) for these studies in our laboratory. The heating rate during the experiments can be maintained as decided, however, in our experiments we have maintained the heating rate at 20 °C per minute. The peak of the DTA curve is recorded as the temperature of Ignition. Weight loss over the temperature interval at pyrolysis and ignition are also noted Thus, from a research perspective, various authors have contributed

Table 11 Thermal and combustion properties of different black liquors

Sample	Bagasse black liquor	Rice straw black liquor	Wheat straw black liquor	Hard wood mixed
Carbon, %	38.87	36.25	37.4	35.6
Organic, %	70.1	79.2	78.8	70.2
Volatiles, % at 400 °C	31	35	35	-
IPDT, °C	574	588	574	430
Ignition temp (Tig), °C	722	662	618	622
SVR, ml/g	15.5	6.2	12	23.0
Cal. value, kJ/g	14.6	10.1	11.6	12.5

Source CPPRI database

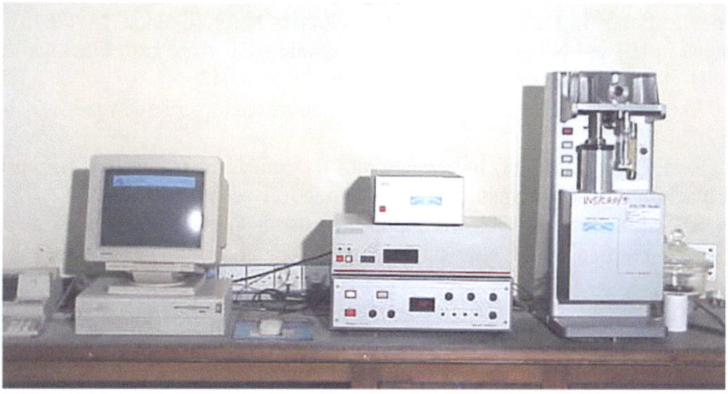

Fig. 4 Photograph of simultaneous thermal analyser (STA-781)

significantly, e.g. in the fields of Kraft black liquor recovery, black liquor viscosity control, separation of alkali from silica-rich black liquor [49–51] in their efforts to find solutions to address the problems being faced by the mill management and the technical shopfloor personnel.

6 Conclusions

Black liquor properties play an important role in the operation of the chemical recovery system; therefore, the characterization of black liquor is crucial to efficiently operate the recovery unit operations. The contributions from different research experts have been highlighted with relevance to the practical problems faced in day to day operations in the mills. Major black liquor properties have been discussed with importance in this chapter with the aim to present a better understanding for the technical personnel working in chemical recovery operations.

References

1. Cardoso M, de Oliveira ÉD, Passos ML (2009) Chemical composition and physical properties of black liquors and their effects on liquor recovery operation in Brazilian pulp mills. Fuel 88(4):756–763. https://doi.org/10.1016/j.fuel.2008.10.016
2. Soderhjelm L (1988) Factors affecting the viscosity of strong black liquor. Appita J 41:389–392
3. Bajpai P (2017) Recovery of chemicals from nonwood black liquor. In: Pulp and paper industry. https://doi.org/10.1016/b978-0-12-811103-1.00007-3
4. Haddad M, Labrecque R, Bazinet L, Savadogo O, Paris J (2016) Effect of process variables on the performance of electrochemical acidification of kraft black liquor by electrodialysis with bipolar membrane. Chem Eng J 304:977–985. https://doi.org/10.1016/j.cej.2016.07.030

5. Manttari M, Lahti J, Hatakka H, Louhi-Kultanen M, Kallioinen M (2015) Separation phenomena in UF and NF in the recovery of organic acids from kraft black liquor. J Membr Sci 490:84–91. https://doi.org/10.1016/j.memsci.2015.04.048
6. Macleod M (2007) Higher liquor solids, steam pressure characterise new recovery boilers. Pulp Pap 81(9):34–5, 37
7. Shen Q, Rosenholm JB (1998) Investigation of the fundamental physico-chemical properties of kraft black liquor. In: International chemical recovery conference, vol 3
8. Hashim SNAS, Zakaria S, Chia CH, Pua FL, Jaafar SNS (2016) Chemical and thermal properties of purified kenaf core and oil palm empty fruit bunch lignin. Sains Malays 45(11):1649
9. Mousavioun P, Doherty WOS (2010) Chemical and thermal properties of fractionated bagasse soda lignin. Ind Crops Prod 31(1):52. https://doi.org/10.1016/j.indcrop.2009.09.001
10. TAPPI T 650 (1971) Solids Content of Black Liquor (Suggested Method). Atlanta, GA: Technical Association of the Pulp and Paper Industry
11. Chai XS, Hou QX, Verrill CL (2004) Rapid determination of total dissolved solids in black liquors by ATR-UV/VIS spectroscopy. In: 2004 international chemical recovery conference, vol 2
12. Sun R, Tomkinson J, Bolton J (1999) Effects of precipitation pH on the physico-chemical properties of the lignins isolated from the black liquor of oil palm empty fruit bunch fibre pulping. Polym Degrad Stab 63(2):195–200. https://doi.org/10.1016/s0141-3910(98)00091-3
13. Milanova E, Dorris GM (1994) On the determination of residual active alkali in black liquors. Nordic Pulp Pap Res J 9:1, 4–9, 15
14. Sullivan J, Radiotis T, Douek M (2007) Improved methods for measuring residual effective alkali in kraft black liquors part 1: absolute measurement by capillary electrophoresis. In: Engineering, pulping and environmental conference 2007, TAPPI, vol 4, pp 2349–2387
15. Radiotis T, Sullivan J, MacLeod M, Syed S, O'Hagan T (2007) Improved methods for measuring residual effective alkali in kraft black liquors, part 2: comparison of titration methods. In: Engineering, pulping and environmental conference 2007, TAPPI, vol 4, pp 2388–2436
16. Ulmgren P, Lindstrom R and Printz M (1994), The alkalinity of black liquor. Nordic Pulp Pap Res J 9(2):76–83, 135
17. Dixit AK, Jain RK, Mathur RM (2012) Innovations & process development for efficient operation of chemical recovery system in paper industry. IPPTA 24(4):119–123
18. Marimuthu P, Reddy KP, Mohan Rao NR (1992) Effect of residual active alkali on viscosity of bagasse kraft black liquor in comparison to wood black liquors. IPPTA 4(3):70–76
19. Silva RV, Cardoso GV, Silva Junior FG, Stanerlin DM, Melo RR, Coeho MU, Pinenta AS (2022) Production of kraft pulp from *Ochroma pyramidale* wood. Acta Amaz 52:81–87
20. Analysis of Sulfate Black Liquor, Tentative Standard, T-625 approved by American National Standard, TAPPI, Atlanta (1964)
21. Andrews JD, Alexander EG, Hart PW (2015) Reconciling material balances with laboratory test results: The case of the inorganic-to-organic ratio in black liquor, Tappi Journal 14(9):593–597
22. Su Z-H, Mahmoud MS, Fan S, Zang Y, Peng F (2019) Combustion properties of mixed black liquor solids from linter and reed pulping. Bio Res 14(4):8278–8288
23. Grace TM, Sachs DG, Grady HJ (1977) Determination of the inorganic composition of alkaline black liquors. Tappi J 60(4):122–125
24. Gupta A, Mathur RM, Kulkarni AG (2001) Non-process elements in non-wood black liquors: their effect on the operation and energy efficiency of chemical recovery systems. Appita J 54(6)
25. Hamaguchi M, Vakkilainen EK (2011) Influence of chlorine and potassium on operation and design of chemical recovery equipment. Tappi J 10(1):33. https://doi.org/10.32964/tj10.1.33
26. Mandavgane SA, Paradkar GD, Varu J, Pamar R, Subramanian D (2007) Desilication of agro based black liquor and green liquor using jet loop reactor. Indian J Chem Technol 14(6):606
27. Yuan H, Wu S, Yin X, Guo D, Wang G, Zhou Z, Wu C (2016) Effects of silica on chemical recovery in the direct causticization of wheat straw black liquor. Tappi J 15(8):557. https://doi.org/10.32964/tj15.8.557

28. Technical Association of the Pulp and Paper Industry (1964) Analysis of soda and sulfate black liquor: Tentative standard T 625 ts-63 (Corrected ed.). Atlanta, GA: Author
29. Alabi SB, Williamson CJ, Lee J (2010) Viscosity models for New Zealand black liquor at low solids concentrations. Asia-Pac J Chem Eng 5(4):619. https://doi.org/10.1002/apj.444
30. Zaman AA, Fricke AL (1994) Correlations for viscosity of kraft black liquors at low solids concentrations. AIChE J 40(1). https://doi.org/10.1002/aic.690400122
31. Zaman AA, Fricke AL (1995) Viscosity of softwood kraft black liquors at low solids concentrations: effects of solids content, degree of delignification and liquor composition. J Pulp Pap Sci 21(4):J119
32. Zaman AA, Fricke AL (1995) Viscoelastic properties of high solids softwood kraft black liquors. Ind Eng Chem Res 34(1):382. https://doi.org/10.1021/ie00040a042
33. Llamas P, Dominguéz T, Vargas JM, Llamas J, Franco JM, Llamas A (2007) A novel viscosity reducer for kraft process black liquors with a high dry solids content. Chem Eng Process Process Intens 46(3):193. https://doi.org/10.1016/j.cep.2006.06.003
34. Zaman AA, Fricke AL (1995) Steady shear flow properties of high solids softwood kraft black liquors: effects of temperature, solids concentrations, lignin molecular weight and shear rate. Chem Eng Commun 139(1):201. https://doi.org/10.1080/00986449508936404
35. Louhelainen J, Alén R, Zielinski J, Sågfors PE (2002) Effects of oxidative and non-oxidative thermal treatments on the viscosity and chemical composition of softwood kraft black liquor. J Pulp Pap Sci 28(9):285
36. Clay DT, Evaporation principles and black liquor properties. https://www.tappi.org/content/events/08kros/manuscripts/3-1.pdf
37. Determination of Calorific Value of Black liquor, TM V-A2, CPPRI, Laboratory Manual of Testing Procedures (2001)
38. Swelling Volume Ratio of Black Liquor TMIII-A12, CPPRI, Laboratory Manual of Testing Procedures (2001)
39. Bhattacharya PK, Shrinath AS, Kunzru D (1985) Pyrolysis of black liquors. J Chem Technol Biotechnol 35:223
40. Sricharoenchaikul V, Phimolmas V, Frederick WJ, Grace TM (1998) Pyrolysis of kraft black liquor: formation and thermal conversion of volatile products and char. J Pulp Pap Sci 24(2):J43
41. Alén R, Rytkönen S, McKeough P (1995) Thermogravimetric behaviour of black liquors and their organic constituents. J Anal Appl Pyrol 31:1
42. Whitty K, Backman R, Forssén M, Hupa M, Rainio J, Sorvari V (1997) Liquor-to-liquor differences in combustion and gasification processes: pyrolysis behaviour and char reactivity. J. Pulp Pap Sci 23(3):J119
43. Gea G, Murillo MB, Arauzo J (2002) Thermal degradation of alkaline black liquor from straw. thermogravimetric study. Ind Eng Chem Res 41(19):4714–4721. https://doi.org/10.1021/ie020283z
44. Yang Q, Yin X, Wu C, Wu S, Guo D (2012) Thermogravimetric-Fourier transform infrared spectrometric analysis of CO_2 gasification of reed (*Phragmites australis*) kraft black liquor. Bioresour Technol 107:512–516. https://doi.org/10.1016/j.biortech.2011.11.06
45. Zhao Y, Bie R, Lu J, Xiu T (2010) Kinetic study on pyrolysis of NSSC black liquor in a nitrogen atmosphere. Chem Eng Commun 197(7):1033–1047. https://doi.org/10.1080/00986440903359418
46. Leite BS, Andreuccetti MT, Leite SAF et al (2013) TG and DSC analyses of eucalyptus black liquor as alternative methods to estimate solids content. J Therm Anal Calorim 112:1539–1544. https://doi.org/10.1007/s10973-012-2689-4
47. Ramesh S, Chaurasia AS, Mahalingam H, Rao NJ (2013) Kinetics of devolatilization of black liquor droplets in chemical recovery boilers - pyrolysis of dry black liquor solids. Int J Chem Eng Appl 4:1. https://doi.org/10.7763/ijcea.2013.v4.248
48. Leite BS, Andreuccetti MT, Leite SAF, D'Angelo JVH (2012) Investigation of thermal analysis techniques potential to determine solids content in eucalyptus black liquor samples. Chem Eng Trans 29:25. https://doi.org/10.3303/CET1229005

49. Fallavollita JA, Mujumdar AS, Avedesian MM (1987) Kraft black liquor recovery in a fluidized bed: Part I — A review. The Canadian Journal of Chemical Engineering, 65(5):812–817. https://doi.org/10.1002/cjce.5450650515
50. Gagnon P, Roy D, Berksoy EM, Boluk Y (1996) Black liquor viscosity control with a flow additive. In 82nd Annual Meeting Technical Section: Creating an Environment for the Future (Preprints B, pp. B169–B175). Montreal, Canada: Canadian Pulp and Paper Association
51. Mansour OY, Nagaty A, El-Khatib MM (1998) Separation of alkali from silica-rich black liquor. Indian Journal of Chemical Technology, 5(1):7–15

Potential of Synthesis of Polyurethane Foam from Black Liquor

Saakshy, Reena Choudhary, and Rahul Mishra

Abstract Lignin has found its diverse applications and one of the potential utility studied was for making polyurethane foam. Polyurethane is a versatile polymer that belongs to the class of "synthetic polymers known as polyurethanes". It is formed through the reaction of "polyols and diisocyanates". The reaction between these two components forms "urethane linkages", which are responsible for the polymer structure. The addition of lignin to polyurethane foams improves the mechanical and protective properties, making them suitable for use in construction, automotive and other industries. In the scope of this chapter, the utilization of lignin in manufacturing polyurethane foam and the effect of properties of polyurethane foam has been covered. Polyurethane foams are important products for the production of polyurethanes, such as hard, semi-rigid and soft foam types. Rigid foams are chiefly employed in furniture, coatings, adhesives and other applications. Flexible and semi flexible foams are extensively used in the automotive sector for seating cushions, bumpers and sound proofing properties. Rigid polyurethane foam with lignin addition provides high insulation and mechanical properties for the building, automotive industry, freeze industry, and naval applications with improved moisture and flame resistivity.

Keywords Lignin · Polyurethane foam · Polyol · Castor oil

1 Introduction

Lignin emerges as a promising renewable alternative to the resources used in fossil-based products, presenting an appealing solution to address the challenge of global climate change generating from the overconsumption of fossil fuels and prolonged emissions of greenhouse gases [6, 23, 34]. In 1838, Anselme Payen, a French chemist, conducted groundbreaking experiments by subjecting wood to nitric acid

---

Saakshy (✉) · R. Choudhary · R. Mishra
Kumarappa National Handmade Paper Institute, Jaipur, India
e-mail: environment.knhpi@gmail.com

S. Sundaramurthy et al. (eds.), *Black Liquor Waste Management in the Pulp and Paper Industry*, Sustainable Materials and Technology,
https://doi.org/10.1007/978-981-96-4237-3_2

and sodium hydroxide, resulting in the isolation of two distinct substances. One of these substances, identified as cellulose, was obtained alongsidc a sccond material rich in carbon content, referred to as "encrusting material" due to its association with cellulose. Lignin constitutes a category of hydrocarbon polymers characterized by both "aliphatic" and "aromatic" structures. Lignin has a three-dimensional structure, amorphous polymer and possesses a substantial molecular weight, featuring "phenyl propane units". These units are derived from the combination of "hydroxyl" and "methoxy-substituted phenylpropane units" [18]. The isolated lignin has "hydroxyl groups" and "aliphatic carbons" in addition to the "benzene rings" [14, 16, 17, 41] (Figs. 1 and 2).

The lignocellulosic raw material is the conventional raw material for paper making. Bagasse, wheat straw, ricc straw, and other agricultural biomass are some of the examples of lignocellulosic raw material and used as raw material comprising of cellulose, hemicellulose and lignin to manufacture paper. After pulping, the black liquor is obtained in liquid form of which main constituent is "Lignin". Black liquor is the principal byproduct of the pulp and paper industry and is alkaline, viscous and dark colored aqueous solution. The organic and inorganic components of black liquor is based on cooking conditions and the type of raw material. Conventionally,

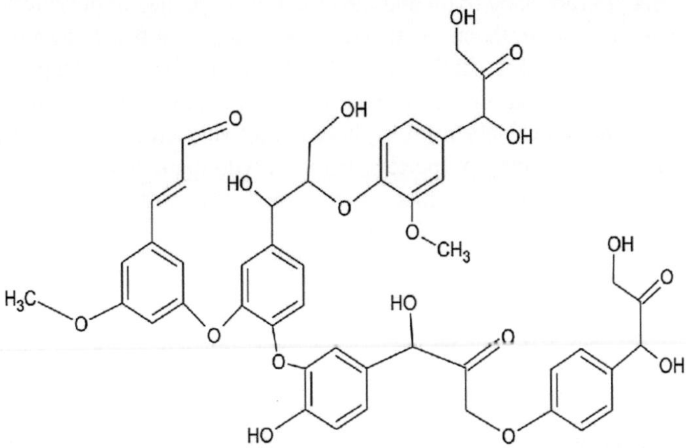

Fig. 1 Chemical structure of lignin

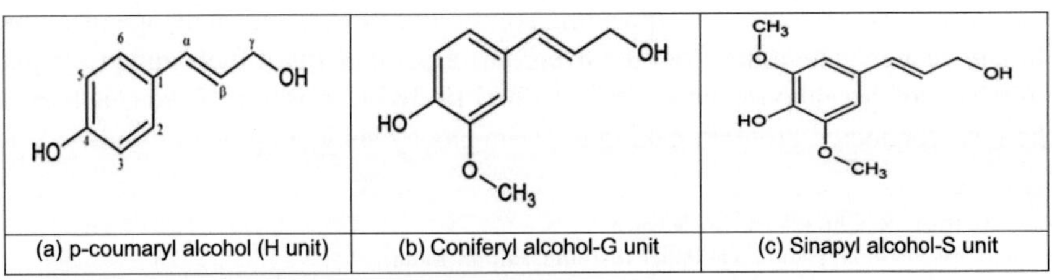

Fig. 2 (a) p-coumaryl alcohol (b) coniferyl alcohol (c) sinapyl alcohol

lignin is burnt for its fuel value in a chemical recovery furnace and only 2% lignin is used for other purposes [27]. There is great potential of utilization of lignocellulosic components to create materials with enhanced added value [2, 30].

The majority of lignin generated as a byproduct in pulp industries is typically incinerated for energy production. While energy is crucial for industrial processes, the practice of burning lignin to generate heat is not only environmentally detrimental but also represents an inefficient utilization of resources [5]. Lignin has various applications including polymers, mixed phenols, mixed aromatic alcohols, mixed organic acids, mixed hydrocarbons, mixed quinones, mixed benzylic aldehydes, aromatics pyrolysis oil, alkyl benzenes, carbon fibers, and activated carbon [51]. Lignin is difficult to mix with different types of polymers due to its high complexity structure and reactivity. Monitoring its behavior and its miscibility with other polymers is crucial to figure out its compatibility and understand the structure property relationship [45].

2 Industrial Aspects of Polyurethane Foam

Polyurethane is a versatile polymer that belongs to the class of synthetic polymers. It is formed by the reaction of compounds with multiple hydroxyl groups (-OH), also called as polyols with compounds having multiple isocyanate groups (-NCO) (di-isocyanates). The reaction between these two components results in the formation of urethane linkage -NH-(C = O)-O- which are responsible for the polymer structure. The physical or chemical blowing agent is used to manufacture polyurethane foams by releasing gas to cause foam to expand [10].

According to Granview research, the global market of polyurethane foam is estimated at USD 43.70 billion in 2023 and expected to grow at compound annual growth rate of 7.8% from 2024 to 2030. The polyurethane foam finds its various applications including bedding and furniture, transportation, seat cushions, head rests, construction, automotive, arm rests, packaging, footwear etc. However, more than 30% of the market of polyurethane foam is for bedding and furniture. Huntsman Corporation, The Dow Chemical company, BASF, Sekisui Chemical are the global top manufacturers of polyurethane foam.

3 Types of Polyurethane Foam

The density, structure, mechanical properties, conductivity and compression are some of the properties of polyurethane foam. It is possible to generate many forms of foam, such as flexible, semi-flexible, or stiff foams, depending on the raw materials, concentrations, and polymerization conditions. These foams possess distinct mechanical resistances and densities, which affect their applications [34].

3.1 Rigid Polyurethane Foam

- Good compressive strength
- High tensile and low elongation properties
- Thermal insulation at temperature from + 130 to -196 °C.
- Lowest thermal conductivity and excellent insulation
- High mechanical properties
- Good density

Uses

- Thermal insulation panels
- Buoyancy shock protective packaging
- Structural reinforcement

Rigid polyurethane foam is low density polyurethane system and used in fire retardants and insulation panels, helmet liners, shoe insoles, plant pots, tractor hoods, electronic enclosures, refrigerated vehicles, road and rail tankers, refrigerated cargos, construction.

3.2 Flexible Polyurethane Foam

- Polyethylene glycol and glycerol is used as the liquefying agent. The tensile strength of flexible polyurethane foam is less than 1 kg/cm^2.
- Compressive strength - It is high desired value for rigid polyurethane foam.
- Compressive strength - Not more than 10% in flexible polyurethane foam.
- The high desired value of 30–200 kg/m^3, while in case of flexible polyurethane foam is not less than 27 kg/m^3. Flexible polyurethane foam is used as an absorbent, padding material and for medical applications [55].

3.3 Production of Flexible Polyurethane Foam from Oil

Castor oil plays a vital role in enhancing the flexibility of the flexible polyurethane foam. The conventional polyols for polyurethane foam preparation are mainly derived from petroleum which prompt interest in renewable alternatives to replace petrochemical-derived compounds [24]. Furthermore, various research studies have documented the replacement of polyols sourced from petrochemical origins with biopolyols derived from vegetable oils in the synthesis of polyurethane [8, 19, 21, 25, 33]. In this context, castor oil has been utilized as a monomer, both in its original form and modified form [4, 8, 47]. The ring-opening reactions and epoxidation [40], thiol-ene coupling in mild conditions [13] to produce hydroxyl-rich biopolyols from vegetable oils; the flexible polyurethane foam is also produced after depolymerization of lignin in the presence of glycerol [51]. The thermal stability of castor oil as replacement of polyol reduces the thermal stability of polyurethane foam [26].

3.4 Production of Rigid Polyurethane Foam from Lignin

Lignin has a complex structure in which the hydroxyl groups and phenolics are entrapped. The high carbon content in lignin acts as a good carbonizing agent to flame retardancy in lignin [32]. Similarly, high aliphatic and phenolic hydroxyl groups in lignin provide good reaction sites with char to form char layer and thus acts as a char forming agent [31]. The oxypropylation process is used to liberate the hydroxyl groups and liquid based lignin polyols are achieved for the synthesis of polyurethane foams [28, 34]. A technology developed for the production of biopolyols involves atmospheric pressure liquefaction [20, 48] (Table 1). In the literature, considerable emphasis has been placed on the preparation of rigid polyurethane foams from lignin-based polymers and the application of lignin-derived polyurethane, as it offers sustainability through the utilization of a renewable byproduct, diminishes dependence on fossil fuels, and provides improved mechanical strength and thermal stability owing to lignin's rigid structure [9, 55].

Table 1 Literature review

S.N.	Title of study	Properties of foam	Lignin preparation method	References
1	"Polyurethane rigid foam, a proven thermal insulating material for applications between +130 and −196 °C"	Rigid foams having lowest thermal conductivity, high mechanical properties at high and low temperature	Rigid polyurethane foam	[11]
2	"Rigid polyurethane foams from lignin based polyols"	Density, mechanical properties, conductivity	Sarkanda lignins with oxypropylated process	[9]
3	"Kraft Lignin based rigid polyurethane foam"	Optimum compressive property achieved by total replacement of Kraft pine lignin	Kraft pine lignin oxypropylation for making polyol	[29]

(continued)

Table 1 (continued)

S.N.	Title of study	Properties of foam	Lignin preparation method	References
4	"Effect of replacing polyol by organosolv and kraft ligin on the property and structure of rigid polyurethane foam"	The lignin replacement ratio has an effect on the lignin-based polyurethane foam density and compressive strength	Hardwood ethanol organosolv lignin and hardwood kraft lignin as replacement of petroleum based polyols up to 30 and 23%, respectively, for making flexible polyurethane foams	[39]
5	"Ligin obtained by acidification at various pH"	Lignin precipitiation with acid	More regularity of cellular structure	[43]
6	"Preparation of bio-based rigid polyurethae foam using hydrolytically depolymerized kraft lignin"	Lignin precipitation with sulphuric acid	Rigid polyurethane foam Kraft lignin by oxypropylaton	[34]
7	"Highly resilient lignin containing polyurethane foam"	Lignin up to 33%, an increase in compressive strain, resilience and density of rigid polyurethane foam	Alkali lignin as partial substitute of petroleum polyols	[51]
8	"Replacement of petroleum-derived diols by sustainable biopolyols in one component polyurethane foams"	Liquefaction of cork powder and eucalyptus bark	Rigid polyurethane foam exhibited high output (16 g/s)	[48]
9	"Highly resilient lignin containing polyurethane foam"	High resilient (elastic recovery more than 93%) and compressive strength increased after the addition of lignin in flexible polyurethane foam	Alkali lignin as a substitute of petroleium polyol	[51]
10	"Rigid polyurethane foams containing lignin oxyalkylated with ethylene carbonate and polyethylene glycol"	Rigid polyurehthane foam having compressive strength up to 204 kPa and density from 51 to 60 kg/m^3	Kraft lignin up to 50%. Oxylalkylation of lignin	[54]

(continued)

Table 1 (continued)

S.N.	Title of study	Properties of foam	Lignin preparation method	References
11	"Lignin as bio-polyol in rigid lightweight wood polyurethane composite foams"	30% wood particles incorporated. Lignin content up to 20%	Rigid lightweight wood polyurethane composite foams. Density of foam increased from 47 to 96 kg/m^3	[32]
12	"Synthesis of lignin based polyurethane foam extracted from black liquor waste"	Rigid polyurethane foam having a blowability of 119.1%, and lowest density of 0.41 g/cm^3	Lignin up to 40%	[36]
13	"Lignin based polyurethane foam extracted from black liquor waste"	Acidification with sulphuric acid	High blowability and low density	[37]
14	"Kraft lignin with improved homogeneity recoverd directly from black liquor and its application in flexible polyurethane foam"s	20% of conventional polyol replaced by lignin	Kraft lignin prepared by acid catalyzed liquefaction	[42]
15	"Rigid polyurethane foam based on kraft lignin"	Oxyalkylation of kraft lignin with polypropylene carbonate	Thermal insulation	[49]
16	"Polyurethane foam made from Kraft lignin related the lignin molecular structure to the efficiency of the rigid polyurethane foams "	Sulphate lignin obtained after acidification	Flexible polyurethan foam developed	[42]
17	Development of rigid polyurethane foams based on kraft lignin polyols obtained by oxyalkylation using propylene carbonate	Rigid polyurethane foam for thermal insulation	Kraft lignin with oxyalkylation process	[49]
18	"Rigid polyurethane from lignopolyol obtained through mild oxypropylation"	Rigid polyurethane foam with good compressive strength	Oxypropylation	[52]

(continued)

Table 1 (continued)

S.N.	Title of study	Properties of foam	Lignin preparation method	References
19	"Diethyl ether as blowing agent, Tween 85 as surfactant, Sulphuric acid as catalyst, Ratio of furfuryl alcohol and lignin"	Kraft lignin	Self blowing lignin furanic foams produced comprising kraft lignin	[15]
20	"Rigid polyurethane foam synthesized using lignin based polyol"	Oxyalkylation of kraft lignin with polypropylene carbonate	Improvement in fire resistance	[50]
21	"Lignin up to 20% and castor oil up to 100% both used as polyols for making polyurethane foams"	Thermal stability improved after the addition of lignin	Rigid polyurethane foam using kraft lignin. Thermal stability of lignin based foam increased while a decrease was observed in castor oil incorporated polyurethane foam	[26]

4 Conclusions

Polyurethane foams stand out as crucial products within the commercial realm of polyurethane and typically categorized as rigid, semi-rigid, or flexible. The main field where polyurethane foams find extensive use is the furniture industry, coating, adhesives, elastomers, cross linker, UV absorber, light stabilizer, antioxidant, flame retardant, primarily utilizing rigid foams. The automotive industry incorporates flexible and semi-flexible polyurethane foam in applications such as seat cushioning, bumpers, sound insulation etc. The rigid polyurethane foam after addition of lignin provides high insulation and mechanical properties which finds its application in construction, automotive, freeze industry and nautical applications. The rigid polyurethane foam provides high insulation, dimensional stability, accelerated ageing and after addition of lignin showed better moisute and flame resistance.

References

1. Alinejad M, Henry C, Nikafshar S, Gondaliya A, Bagheri S, Chen N, Singh SK, Hodge DB, Nejad M (2019) Lignin-based polyurethanes: opportunities for bio-based foams elastomers, coatings and adhesives. Polymers 11:1202
2. Asina FNU, Brzonova I, Kozliak E, Kubátová A, Ji Y (2017) Microbial treatment of industrial lignin: successes, problems and challenges. Renew Sustain Energy Rev 77:1179–1205
3. Bernardini J, Cinelli P, Anguillesi I, Coltelli MB, Lazzeri A (2015) Flexible polyurethane foams green production employing lignin or oxypropylated lignin. Eur Polym J 64:147–156

4. Bresolin D, Valério A, de Oliveira D, Lenzi MK, Sayer C, de Araújo PHH (2018) Polyurethane foams based on biopolyols from castor oil and glycerol. J Polym Environ 26:2467–2475
5. Bruijnincx PC, Weckhuysen BM (2014) Lignin up for break-down. Nat Chem 6(12):1035–1036
6. Cao Y, Chen SS, Zhang S, Ok YS, Matsagar BM, Wu KCW, Tsang DC (2019) Advances in lignin valorization towards bio-based chemicals and fuels: lignin biorefinery. Bioresour Technol 291:121878
7. Cao H, Liu R, Li B, Wu Y, Wang K, Yang Y, Li A, Zhuang Y, Cai D, Qin P (2022) Biobased rigid polyurethane foam using gradient acid precipitated lignin from the black liquor: revealing the relationship between lignin structural features and polyurethane performances. Ind Crops Prod 177:114480
8. Carriço CS, Fraga T, Pasa VM (2016) Production and characterization of polyurethane foams from a simple mixture of castor oil, crude glycerol and untreated lignin as bio-based polyols. Eur Polym J 85:53–61
9. Cateto C, Barreiro F, Rodrigues A, Belgacem N (2008) Rigid polyurethane foams from lignin-based polyols. In: 4th International Conference on Times of Polymers and Composites, pp 243–245
10. Coste G, Negrell C, Caillol S (2020) From gas release to foam synthesis, the second breath of blowing agents. Eur Polym J 140:110029
11. Demharter A (1998) Polyurethane rigid foam, a proven thermal insulating material of applications between +130 °C and −196 °C. Cryogenics 38:113–117
12. Desouza RE, Gomes FJB, Brito EO, Lelis RCC, Batalha LAR, Santos FA, Junior DL (2020) A review on lignin sources and uses. J Appl Biotechnol Bioeng 7(3):100–105
13. Desroches M, Caillol S, Lapinte V, Auvergne R, Boutevin B (2011) Synthesis of biobased polyols by thiol−ene coupling from vegetable oils. Macromolecules 44(8):2489–2500
14. Ekeberg D, Gretland KS, Gustafsson J, Braten SM, Fredheim GE (2006) Characterization of lignosulphonates and kraft lignin by hydrophobic interaction chromatography. Analytica Chemica Acta 565(1):121–128
15. Fagbemigun TW, Mail C (2023) Production and characteristics of self-blowing lignin-based foams. Eur J Wood Wood Prod 81:579–590
16. Fodil Cherif M, Trache D, Brosse, Benaliouche F, Tarchaun AF (2020) Comparison of physicochemical properties and thermal stability of organosolv and kraft lignins from hardwood and softwood biomass for their potential valorization. Waste Biomass Valori 11:6541–6553
17. Garlapati VK, Chandel AK, Kumar SJ, Sharma S, Sevda S, Ingle AP, Pant D (2020) Circular economy aspects of lignin: towards a lignocellulose biorefinery. Renew Sustain Energy Rev 130:109977
18. Ghalia MA, Dahman Y (2017) Synthesis and utilization of natural fiber-reinforced poly (lactic acid) bionanocomposites. In: Lignocellulosic fibre and biomass-based composite materials. Woodhead Publishing, pp 313–345
19. Ghasemlou M, Daver F, Ivanova EP, Adhikari B (2019) Polyurethanes from seed oil-based polyols: a review of synthesis, mechanical and thermal properties. Ind Crops Prod 142:111841
20. Gurgel D, Bresolin D, Sayer C, Filho LC, Hermes de Araújo PH (2021) Flexible polyurethane foams produced from industrial residues and castor oil. Ind Crops Prod
21. Hejna A, Kirpluks M, Kosmela P, Cabulis U, Haponiuk J, Piszczyk Ł (2017) The influence of crude glycerol and castor oil-based polyol on the structure and performance of rigid polyurethane-polyisocyanurate foams. Ind Crops Prod 95:113–125
22. Hormaiztegui MV, Aranguren MI, Mucci VL (2018) Synthesis and characterization of a waterborne polyurethane made from castor oil and tartaric acid. Eur Polym J 102:151–160
23. Hu J, Zhang Q, Lee DJ (2018) Kraft lignin biorefinery: a perspective. Biores Technol 247:1181–1183
24. Huo S, Jin C, Liu G, Chen J, Wu G, Kong Z (2019) Preparation and properties of biobased autocatalytic polyols and their polyurethane foams. Polym Degrad Stab 159:62–69
25. Jo YJ, Ly HV, Kim J, Kim SS, Lee E (2015) Preparation of biopolyol by liquefaction of palm kernel cake using PEG# 400 blended glycerol. J Ind Eng Chem 29:304–313

26. Kim HJ, Jin X, Choi JW (2024) Investigation of bio-based rigid polyurethane foams synthesized with lignin and castor oil. Sci Rep 14:13490
27. Kumar A, Ray AK, Schable S, Kroline H (2016) Application of lignin as filler for the middle layer of carton board. IPPTA 28:62–70
28. Kühnel I, Saake B, Lehnen R (2017) Oxyalkylation of lignin with propylene carbonate: influence of reaction parameters on the ensuing bio-based polyols. Ind Crops Prod 101:75–83
29. Li Y, Ragauskas AJ (2012) Kraft lignin-based rigid polyurethane foam. J Wood Chem Technol 32(3):210–224
30. Lorenci Woiciechowski A et al (2020) Lignocellulosic biomass: acid and alkaline pretreatments and their effects on biomass recalcitrance – conventional processing and recent advances. Bioresour Technol 304:122848. https://doi.org/10.1016/j.biortech.2020.122848
31. Lu Y, Lu Y-C, Hu H-Q, Xie F-J, Wei X-Y, Fan X (2017) Structural characterization of lignin and its degradation products with spectroscopic methods. J Spectrosc. https://doi.org/10.1155/2017/8951658
32. Luo S, Gao L, Guo W (2020) Effect of incorporation of lignin as bio-polyol on the performance of rigid lightweight wood-polyurethane composite foams. J Wood Sci 66(23):1–10
33. Luo X, Hu S, Zhang X, Li Y (2013) Thermochemical conversion of crude glycerol to biopolyols for the production of polyurethane foams. Bioresour Technol 139:323–329
34. Mahmood N, Yuan Z, Schmidt J, Xu CC (2015) Preparation of bio-based rigid polyurethane foam using hydrolytically depolymerized kraft lignin via direct replacement or oxypropylation. Eur Polym J 68:1–9
35. Mahmood Z, Yameen M, Jahangeer M, Riaz M, Ghaffar A, Javid I (2018) Lignin as natural antioxidant capacity, vol 10. IntechOpen, London, UK, pp 181–205
36. Mardiyati Y, Prabaswara EA, Steven S (2020) Synthesis of lignin-based polyurethane foam extracted from black liquor waste. AIP Conf Proc 2262:060011
37. Mohammadpur R, Sadeghi GMM (2020) Potential use of black liquor as lignin source for synthesis of polyurethane foam. J Polym Res 27:362
38. Murmu SB (2022) Alternatives derived from renewable natural fibre to replace conventional polyurethane rigid foam insulation. Clean Eng Technol 8:100513
39. Pan X, Saddler JN (2013) Effect of replacing polyol by organosolv and kraft lignin on the property and structure of rigid polyurethane foam. Biotechnol Biofuels 6:1–10
40. Pan X, Webster DC (2012) New biobased high functionality polyols and their use in polyurethane coatings. ChemSusChem 5(2):419–429
41. Ponnusamy VK, Nguyen DD, Dharmaraja J, Shobana S, Banu JR, Saratale RG, Chang SW, Kumar G (2019) A review on lignin structure, pretreatments, fermentation reactions and biorefinery potential. Bioresour Technol 271:462–472
42. Quan P, Kiziltas A, Gondaliya A, Siahkamari M, Nejad M, Xie X (2022) Kraft lignin with improved homogeneity recovered directly from black liquor and its application in flexible polyurethane foams. ACS Omega 7:16705–16715
43. Radoykova T, Nenkova S, Valchev I (2013) Black liquor lignin products, isolation and characterization. J Chem Technol Metall 48(5):524–529
44. Ragauskas AJ, Beckham GT, Biddy MJ, Chandra R, Chen F, Davis MF, Davison BH, Dixon RA, Gilna P, Keller M, Langan P (2014) Lignin valorization: improving lignin processing in the biorefinery. Science 344(6185):1246843
45. Sen S, Patil S, Argyropoulos DS (2015) Thermal properties of lignin in copolymers, blends, and composites: a review. Green Chem 17(11):4862–4887
46. Torres LAZ, Woiciechowski AL, de Andrade Tanobe VO, Karp SG, Lorenci LCG, Faulds C, Soccol CR (2020) Lignin as a potential source of high-added value compounds: a review. J Clean Prod 263:121499
47. Trovati G, Suman MVN, Sanches EA, Campelo PH, Neto RB, Neto SC, Trovati LR (2019) Production and characterization of polyurethane castor oil (*Ricinus communis*) foam for nautical fender. Polym Test 73:87–93
48. Vale M, Mateus MM, dos Santos RG, de Castro CN, De Schrijver A, Bordado JC, Marques AC (2019) Replacement of petroleum-derived diols by sustainable biopolyols in one component polyurethane foams. J Clean Prod 212:1036–1043

49. Viera FR, Gama NV, Barros-Timmons A, Evtuguin DV, Pinto PCOR (2022) Development of rigid polyurethane foams based on kraft lignin polyol obtained by oxyalkylation using propylene carbonate. Chem Eng 6:95
50. Viera FR, Gama NV, Evtuguin DV, Amorim CO, Amaral VS, Pinto PCOR, Barros-Timmons A (2023) Bio-based polyurethane foams from kraft lignin with improved fire resistance. Polymers 15(1074):1–18
51. Wang S, Liu W, Yang D, Qiu X (2019) Highly resilient lignin-containing polyurethane foam. Ind Eng Chem Res 58(1):496–504
52. Wu M, Peng J-J, Dong Y-M, Pang J-H, Zhang X-M (2022) Preparation of rigid polyurethane foam from lignopolyol obtained through mild oxypropylation. R Soc Chem RSC Adv 12:21736–21741
53. Zhang X, Jeremic D, Kim Y, Street J, Shmulsky R (2018) Effects of surface functionalization of lignin on synthesis and properties of rigid bio-based polyurethanes foams. Polymers 10:706
54. Zhang X, Kim Y, Elsayed I, Taylor M, Eberhardt TL, Hassan EB, Shmulsky R (2019) Rigid polyurethane foams containing lignin oxyalkylated with ethylene carbonate and polyethylene glycol. Ind Crops Prod 141:111797
55. Zhang Q, Zhang G, Xu J, Gao C, Wu Y (2015) Recent advances on ligin-derived polyurethane polymers. Rev Adv Mater Sci 40(2):146–154

Trends in Analytical Techniques for Lignin Quantification and Characterization

Saakshy and Rahul Mishra

Abstract Lignin is another most abundantly available organic compound/biopolymer after cellulose and greatest source of natural aromatic compounds on earth. Lignin comes under the category of organic polymer and play an important role in the formation of "cell walls" in wood and lend "rigidity to the cell walls". In the "pulp and paper sector", lignin is present in the "black liquor" which is received after the pulping or delignification process. There is increasing need of extraction, purification and characterization of the lignin to explore its application in various industries. The analytical techniques of lignin is important and therefore, the quantitative and qualitative analytical methods of lignin is required to be explored for better utilization of lignin in various industrial applications. In the scope of this chapter, the details of analytical techniques of lignin has been mentioned including the characterization of chemical structure, particle size and molecular size determination.

Keywords Lignin · Quantitative estimation · Structural characterization · Particle size · Molecular weight · Thermal analysis

1 Introduction

The value-added products such as surfactants and dispersants, thermoplastics, resins, composites, and different fine chemicals like vanillin made from lignin is becoming more widely known [35, 40, 51]. However, lignin need to be characterized in order to determine its suitability for use in the production of a particular product. The characterization and composition of lignin depends on the "source of raw material", "pulping process", "extraction process", "purity" and vary accordingly. The "global market size of lignin was estimated at USD 954.5 million in 2019 and expected to expand at an annual rate of 2% from 2020 to 2027" (corona-effects-on-business.blogspot.com; grandviewresearch.com). In the growing demand market

Saakshy (✉) · R. Mishra
Kumarappa National Handmade Paper Institute, Jaipur, India
e-mail: saakshyagarwal@gmail.com

S. Sundaramurthy et al. (eds.), *Black Liquor Waste Management in the Pulp and Paper Industry*, Sustainable Materials and Technology,
https://doi.org/10.1007/978-981-96-4237-3_3

of lignin, the market of "concrete admixtures, adhesives and binders coupled with increasing investments in infrastructure development is expected to fuel the market growth" (corona-effects-on-business. blogspot.com) [12]. There are various parameters of lignin including characterization of chemical structure, molecular weight, particle size and thermal properties which needs to be analyzed using the latest equipment and technologies.

2 Quantitative Estimation of Lignin

The total lignin is summation of "acid soluble lignin" and "acid insoluble lignin".

2.1 *Insoluble Lignin*

Peter Klason in 1923 developed the Klason method for estimation of insoluble lignin. He used "sulphuric acid in concentration of 61–68 wt. % at low temperature to obtain an acid-insoluble lignin". After 10 years, Ritter increased the concentration of sulphuric acid to "72%" and used it for the analysis of insoluble lignin. In [29], Nicholson further modified the method and got more accurate results [29]. Still, the conventional "Klason method of insoluble lignin determination" is most broadly utilized method [24, 28, 39].

2.2 *Soluble Lignin*

The soluble lignin is estimated with the help of "UV–Vis spectroscopy" by scanning the sample in solution form after centrifugation in the UV range. The lignin is aromatic in nature which gets absorbed in ultra violet range. The maxima is determined in UV range and the quantitative estimation of sample is made with the corresponding linear graph prepared with standard lignin.

3 Structural Characterization of Lignin

The structure of the lignin contains "three phenols - "coniferyl (G lignin)", "sinapyl alcohol (S lignin)", and "p-coumaryl (H lignin)" which are "cross-linked by carbon–carbon and oxygen bonds which make lignin, a branching polymer" [10].

A "1, 4-substituted aromatic ring with an OH group on one side and a $-CH = CH_2OH$ on the other" is the fundamental unit of lignin. "Coniferyl lignin has one methoxyl group on one side of the OH, whereas Sinapyl lignin has two methoxyl groups on either side of the OH" [18], https://www.spectroscopyonline.com). The structure of lignin is characterized by "Scanning Electron Microscopy", "Transmission Electron Microscopy" and "Atomic Force Microscopy" (worldwidescience.org).

3.1 *Electron Microscopy*

Scanning electron microscopy (SEM) has been widely utilized to examine lignin "surface dispersion patterns, morphological changes, and microstructure" [50]. SEM is used to observe lignin droplets in size from roughly "0.05–2 mm" [4]. Field-based emission SEM is typically used to provide fractures with great resolution [34]. "Transmission electron microscopy" (TEM) is used to evaluate the "ultra-structure and distribution of lignin in plant cell walls" [49].

3.2 *Atomic Force Microscopy*

The molecular conformation of lignin is characterized by the use of atomic force microscopy (AFM) [16], worldwidescience.org). It is capable of giving precise lignin distribution patterns [13]. In 2018, Lambert developed dynamic in situ real-time Atomic Force Microscopy, which allowed him to view the distribution of lignin and study ligno-cellulosic biomass [14, 22]. This technique can determine the microfibrills in S1 and S2 layer.

4 Chemical Structure Determination of Lignin

The characterization of the chemical structure of lignin can be conducted by UV spectroscopy [25], "FTIR spectroscopy", "Raman spectroscopy", "Nuclear Magnetic Response spectroscopy" and "X-ray photoelectron microscopy". The spectroscopy methods detect the lignin moieties directly and for structural elucidation both in solid state and in liquid form.

4.1 *UV Spectroscopy*

The UV-spectroscopy method is adopted when lignin is dispersed in a suitable solvent (water, dichloro ethane, 1, 4 dioxane, hexafluoropropanol, pyridine, dimethylsulfoxide, 2-methoxyethanol, dimethylformamide, ethanol) or in the form of black liquor [41]. In some cases, the combination of solvents are also used for structural determination of lignin. The combination of solvents include DMSO/water, DMSO/

ethanol, ethylene glycol, propylene carbonate/ethanol/water [38]. The maxima of the lignin is determined by scanning the sample in the "UV range of 200–400 nm". The maxima at different wavelength in UV range shows the presence of various groups. The maxima at different wavelength showes the presence of "methoxylated phenolic ring at 220 nm"; "biphenyl derivatives at 240 nm"; "280 m shows the presence of benzene ring substituted by methoxyl or hydroxyl groups"; "free and etherified hydroxyl group at 298 nm"; "aromatic carboxylic acids and carbonyl groups at 320 nm"; "carbonyl groups", "double bond conjugated phenols" and "dissociated aromatic carboxylic acids at 365 nm" [32].

4.2 Fourier Transform Infrared Spectroscopy

FTIR spectroscopy is a powerful method for determining the "functional groups" present at the surface of the sample. FTIR spectra of the lignin samples is recorded on "KBr pellets" with a spectral resolution of "4 cm^{-1} with frequency range of 400–4600 cm^{-1}" [27]. The IR spectra is recorded after an interval of specific time during both heating and cooling. The assignment of bands at various wavenumber (cm^{-1}) show the presence of various functional groups as per its assignment bands [9].

4.3 Raman Spectroscopy

The Raman spectroscopy is a semi-quantitative technique where an internal standard normalize the spectra for the comparison of the intensities [1]. In Raman spectroscopy, the sample is excited when it is irradiated by a laser beam. The photon frequency varies as opposed to reporting the photons absorbed at a fixed frequency and this shift is a result of the "Raman effect". The difference in energy between the "original state and the post- excitation ground state" shifts the photon's frequency [21]. Table 1 shows the functional groups at various wavenumbers.

4.4 Nuclear Magnetic Resonance Spectroscopy

In order to determine the composition of "H/G/S units, monolignols, and the frequency of links between them in lignin", the NMR approach yields high-resolution qualitative and quantitative data. The structural and compositional studies of lignin have been conducted using "^{1}H NMR, ^{13}C NMR, ^{19}F NMR, ^{31}P NMR, and several modes of 2D NMR" techniques. Both 2D heteronuclear single-quantum coherence (HSQC) NMR and 13C NMR offer precise quantitative data on the side-chain moieties and functional groups of both liquid and solid lignin. Employing "1H

Table 1 Assignment of bands in Raman spectroscopy [44]

Assignments	Wavenumbers, cm^{-1}	Functional groups
O–H stretching	3400–3600	Free hydroxyl group
O–H stretching	3100–3400	Associated hydroxyl group
O–H stretching	2920	Carboxylic hydroxyl group
C–H stretching	2820–2960	$-CH_2$, $-CH_3$
C–H stretching	2650–2890	Methyl group in methoxyl
C–H stretching	1140	Guaiacyl
C–H stretching	1130	Syringyl
C–H stretching	750–860	Aromatic ring
C=O stretching	1771	Aromatic carbonyl
C=O stretching	1700–1750	Ketones, esters and carbonyl groups
C=O stretching	1722	Aliphatic carbonyl
C=O stretching	1630–1680	Conjugated p-substituent carbonyl and carboxyl
C–O stretching	1325–1330	Syringyl ring
C–O stretching	1230–1235	
C–O stretching	1088	Aliphatic ether and secondary alcohol

(500 MHz) and 13C (125 MHz) NMR spectroscopy in acetone-d6", the purity of lignin is determined.

^{1}H NMR

The structure of lignin is investigated using ^{1}H NMR, which also determines the proton's chemical environment in "H-containing moieties such as aliphatic and aromatic hydroxyl, benzylic hydroxyl and aliphatic hydrogen atoms". The protons within the anhydroxylose units are responsible for the signals detected at "3–10 ppm" [23]. The chemical alterations in the "3.3–4.0 ppm" range are caused by "protons in methoxyl groups". The peaks in the "4.0–6.3 ppm" range are associated with aliphatic protons [13] in the pathways "β–O–4, β–β, and β–5". The "aromatic protons in H, G, and S units" are responsible for the chemical shifts at approximately "7.5, 7.0, and 6.5 ppm", respectively [5]. Derivatization techniques are typically used to quantify the "H-containing functional groups" in lignin [20] because of the limited "range of chemical shifts" significant "signal overlapping", and impacts of "proton coupling" [3].

4.5 *X-ray Photoelectron Microscopy*

The "surface chemistry, degree of disorder, bonding states, and relative contents of atoms in poly-aromatic stacks" can be characterized using X-ray photoelectron spectroscopy (XPS) [29]. Semi-quantitative estimation of the lignin content can be achieved by analyzing the elemental compositions, the "aliphatic carbon component", and the "oxygen–carbon ratio" [19, 42]. High cellulose and hemicellulose content is indicated by an increased O/C ratio, whereas higher lignin concentration is suggested by a lower O/C ratio. The "lignification and polymerization of lignin" result in a sharp rise in "chemical shift" and a fall in C2, C3, and O2, which lowers the O/C ratio [19].

5 Molecular Weight Determination of Lignin

The molecular weight of lignin is determined by the "degree of polymerization". The molecular weight of the lignin can be analyzed with the help of Gel Permeation Chromatography (GPC). The molecular weight of lignin can also be determined with the help of High Performance Liquid Chromatography (HPLC).

5.1 Gas Chromatography (GC)

GC is an effective method that is used to separate volatile substances based on the differences in their structural characteristics and boiling points. GC focuses on finding volatile chemical compounds with molecular weights between "50 and 500 Da", "lower polarity, and lower boiling points (<350 °C)". Lignin-related complex materials have been extensively characterized using GC in conjunction with a variety of detectors [17, 49]. Certain functional groupings are sensitive to different detectors' resolutions and sensitivities, each of which has its own intrinsic limitations. For a single lignin sample, hundreds of species can be recognized concurrently in a single run. The assignment of species in a chromatogram is one of the difficulties in GC analysis. The sample's species may only have partial characterization. For instance, the overlapping of peaks and the heavy reliance on the database for component identification complicate the total ion chromatogram that results from GC/MS analysis. Py-GC/MS has been widely utilized in recent years and has shown to be an effective method for lignin analysis, including content, "molecular weight distribution", "composition of monomers and oligomers", "linkage frequency", and "thermodynamic" characteristics [26].

5.2 High Performance Liquid Chromatography (HPLC)

High performance liquid chromatography (HPLC) is an analytical technique useful to determine the "ratio of the lignin monomers" obtained by the "alkaline nitrobenzene oxidation method" [37]. The physical and spectral properties of aromatic fragment of lignin has been analyzed by HPLC [2]. The lignin components travel at varying speeds when the sample passes through the sorbent-filled column, depending on their chemical composition. An high performance liquid chromatography (HPLC) method, "applying diode array (DAD) and coulometric array detection simultaneously, was developed to determine monomeric phenolic compounds in lignin hydrolysates" [11]. The pure lignin shows low dispersity and high molecular weight. The purity was also checked by HPLC–MS, the separation was done using column and acetonitrile gradient using 0.1% formic acid.

6 Particle Size Determination of Lignin

Lignin particles can range in size from 50 to 200 μm, with most particles being around 76 μm in size. The particles have a thickness of "1–3 nm and a length of 5–9 nm", which are dependent on the solvent and lignin concentration. Lignin particle size distribution is a key parameter in lignin production as it impacts lignin cake filtration and washing rates, as well as several lignin applications such as thermoplastics and asphalts in which lignin is blended with other polymers (e.g. polypropylene) (standardsdevelopment.bsigroup.com) and bitumen, respectively. Use of lignin for producing these and other related materials will be of great benefit in reducing their carbon footprint, since lignin will largely replace fossil-based chemicals that are currently used for these applications. The particle size distribution at various image analysis methods is given as Table 2. There are two ways of determination of particle size distribution of lignin i.e. laser light diffraction and dynamic image analysis techniques.

6.1 Laser Light Diffraction

Laser diffraction analysis is the most common method for the determination of "particle size distribution" [33]. The method is based on the "deflection of a laser beam by an ensemble of particles dispersed in either a liquid or an air stream" (www.microtrac.com). A laser light from a monochromatic light source is passed through a lignin in dispersed medium and the light scattered by the lignin particles at various angles depending on the shape and size is measured by an "array of photo detectors". The method is based on the "deflection of a laser beam" by an "ensemble of particles dispersed in either a liquid or an air stream". The "angles of diffraction or scattering angles are characteristic of the particle size". "ISO 13320 comprehensively describes the methodology of laser diffraction".

Table 2 Particle size distribution at various image analysis methods

	PARTICLE SIZE DISTRIBUTION				
	Less than 1nm	1nm to 1µm	1µm to 1mm	1mm to 10cm	10cm and above
Dynamic Image Analysis			Camsizer 2 D 0.8µ to 8mm		
			Camsizer 3 D 20µ to 30mm		
				Camsizer XL 160µm to 135mm	
Laser Diffraction			SYNC 10nm to 4mm		
			AEROTRAC II 0.5 µm to 2mm		
Dynamic light scattering		NANOTRAC SERIES **0.3nm** to 10µm			
Sieve Analysis				20µm to 125mm	

6.2 Dynamic Image Analysis

Using a camera to capture and record an image of scattered particles, "dynamic image analysis" can be used to identify the "shape and particle size distribution of a sample". The images are then analyzed in real time resulting in high resolution. Dynamic Image Analysis (DIA) can determine the particle size distribution based on different size definitions, such as particle width, particle length, or diameter of the equal area circle. The sizes are determined as chord dimensions, Feret lengths, or Martin diameters. Repeated experiment results produced using DIA show a small dispersion, suggesting that a very accurate particle size distribution can be created.

"Dynamic light scattering is based on Brownian motion of particles" in suspensions. In DIA, a camera system detects particles moving through it and records their images as shadow projections. These images are then analyzed in real time, resulting in a high-resolution size and shape distribution within minutes. DIA can also be used to analyze particle shape, such as roundness, circularity, aspect ratio, convexity, and symmetry.

7 Thermal Analysis of Lignin

7.1 Determination of Glass Transition Temperature (Tg)

The glass transition temperature of lignin can be determined by differential scanning calorimetry (DSC). The black liquor solution which is usually alkaline in nature is acidified with the help of sulphuric acid, hydrochloric acid and carbonation process. The sample should be powder in form. The glass transition temperature of lignin is used to determine the softening point.

7.2 Determination of Thermal Stability by Thermogravimetry

Thermogravimetry can be used to determine the initial decomposition temperature(s), rates of decomposiion and the temperature at maximum decomposition.

7.3 Future research prospects

For future research in this field, it is recommended that the structural features of lignin, lignin functionality, particle size distribution for lignin powders, and the aromatic fragments from lignin degradation be evaluated using NMR/HPLC [6–8, 31, 45, 47, 48]. Additionally, the significance of wood in mitigating climate change, wood variability, processing and utilization, wood protection, wood-based composites, and wood energy should be assessed from a policy, governance, and technological perspective [30]. The lignin market size and trends should be analysed and evaluated on a regular basis, at least within the past two years, to facilitate commercialization [36]. Additionally, research should be intensified to identify the acid-insoluble lignin (AIL), also referred to as Klason lignin, or lignin removal enhancement from the pre-hydrolysis liquor, among other things, to improve the production of value-added products, such as bioethanol [43, 46]. The application of green pretreatment technologies, the reduction of biomass recalcitrance, and the valorization of lignin to bioplastics and carbon fibers should be investigated and evaluated at various technology readiness levels (TRLs) from a resource recovery perspective.

8 Conclusions

The characterization of lignin plays a significant role in the determination of suitability of lignin in various end applications. Lignin, being a complex organic polymer, the characterization needs sophisticated equipment with various methods applicable according to the nature of lignin and the required precision. The methods have been regularly modified and the new techniques added with recent studies in order to have more precise characterization of lignin for the better applications in various sectors. The characterization of lignin is also important in understanding chemical reactions and the new compounds formulation with lignin in addition to the degradation of lignin. There are various types of lignocellulosic raw material having different amount of lignin, the extraction processes are different and therefore the type of lignin achieved is entirely different. Threfore, the characterization including chemical structure, molecular weight, thermal properties of lignin utilizing sophisticated equipment can well explain the suitability of lignin emitted from industries as a by product. The complete characterization is helpful in exploring the utilizaiton of lignin as raw material or additive in various industries. Lignin being a renewable source and abundantly available shall be sustainable source of raw matieral after its characterization utilizing latest sophisticated equipment and techniques.

References

1. Agarwal UP (2011) Lignin quantitation by FT-Raman spectroscopy. In: 16th international symposium on wood, fiber and pulping chemistry - proceedings, ISWFPC, 1 January 2011, pp 170–173
2. Ali S, Rani A, Dar MA, Qaisrani MM, Noman M, Yoganathan K, Asad M, Beshanu A, Barwant M, Zhu D (2024) Recent advances in characterization and valorization of lignin and its value added products: challenges and future perspectives. Biomass 4(3):947–977
3. Balakshin M, Capanema EA, Santos RB, Chang HM, Jameel H (2016) Structural analysis of hardwood native lignins by quantitative ^{13}C NMR spectroscopy. Holzforschung 70(2):95–108
4. Beisl S, Miltner A, Friedl A (2017) Lignin from micro to nanosize: production methods. Int J Mol Sci 18(6):1244
5. Bhang FSG, Ray S (2022) Lignin extraction from black liquor using acetate based ionic liquids: exrraction kinetics and product characterization. J Ionic Liq 2:100028
6. Capanema EA, Balakshin MY, Kadla JF (2004) A comprehensive approach for quantitative lignin characterization by NMR spectroscopy. J Agric Food Chem 52(7):1850–1860
7. Chen WJ, Zhao BC, Wang YY, Yuan TQ, Wang SF, Sun RC (2018) Revealing the topochemistry and structural features of lignin during the growth of *Eucalyptus grandis* × *Eucalyptus urophylla*. ACS Sustain Chem Eng 6(7):9198–9207
8. Company C, Synthe L (1998) Characterization of lignin functionality by NMR 19:1–9
9. Derkacheva O, Sukhov D (2008) Investigation of lignins by FTIR spectroscopy. Macromol Symp 265(1):61–68
10. Eswaran SCD, Subramaniam S, Sanyal U, Rallo R, Zhang X (2022) Molecular structural dataset of lignin macromolecule elucidating experimental structural compositions. Sci Data 9(1):647
11. Fischer K, Höffler S (2021) RP-HPLC analysis of phenolic lignin monomers with DAD and coulometric array detection. J Chromatogr Sci 59(4):312–323

12. G V Research (2020) Lignin market size, share & trends analysis report by product (Lignosulphonates, kraft, organosolv) by application (macromolecule, aromatic), By Region and Segment Forecasts, 2020–2027
13. Gidh AV, Decker SR, See CH, Himmel ME, Williford CW (2006) Characterization of lignin using multi-angle laser light scattering and atomic force microscopy. Anal Chim Acta 555(2):250–258
14. Hollertz R, Arwin H, Faure B (2013) Dielectric properties of lignin and glucomann as determined by spectroscopic ellipsometry and lifshitz estimates of non-retarded hamaker constants. Cellulose 20:1639–1648
15. Jones RR, Hooper DC, Zhang L, Wolverson D, Valev VK (2012) Raman techniques: fundamentals and frontiers. Nanoscale Res Lett 14(1):231
16. Joshi J, Homburg SV, Ehrmann A (2022) Atomic Force Microscopy (AFM) on biopolymers and hydrogels for biotechnological applications-possibilities and limits. Polymers 14(6):1267
17. Kaiser K, Benner R (2012) Characterization of lignin by gas chromatography and mass spectrometry using a simplified CuO oxidation method. Anal Chem 84(1):459–464
18. Katahira R, Elder TJ, Beckham GT (2018) Chapter 1: A Brief Introduction to Lignin Structure, in: Lignin Valorization: Emerging Approaches, ed. G. T. Beckham, The Royal Society of Chemistry, Energy and Environment Series, pp. 1–20
19. Kaur K, Kaur R, Kaur H (2024) A systematic review of ligno-cellulosic biomass for remediation of environmental pollutants. Appl Surf Sci Adv 19:100547
20. Kenny JK, Medlin JW, Beckham GT (2023) Quantification of phenolic hydroxyl groups in lignin via ^{19}F NMR spectroscopy. ACS Sustain Chem Eng 11:5644–5655
21. Kumar R, Gupta A, Chawla M, Aadil KR, Dutt S, Kumar VB, Chaudhary A (2020) Advances in nanotechnology based strategies for synthesis of nanoparticles of lignin. In: Sharma S, Kumar A (eds) Lignin. Springer Series on Polymer and Composite Materials, Springer, Cham
22. Lambert E, Aguie-Beghin V, Dessaint D, Foulon L, Chabbert B, Paes G, Molinari M (2019) Real time and quantitative imaging of lignocellulosic films hydrolysis by atomic force microscopy reveals lignin recalcitrance at nanoscale. Biomacromolecules 20(1):515–527
23. Li J, Zhang J, Zhang S, Gao Q, Li J, Zhang W (2017) Fast using bio-based phenolic resins via lignin demethylated under mild reaction condition. Polymers 9(9):428
24. Sharma S, Kumar A (Eds.) (2020) Lignin: Biosynthesis and transformation for industrial applications. Springer Nature. https://link.springer.com/book/10.1007/978-3-030-40663-9
25. Lu Y, Lu Y-C, Hu H-Q, Xie F-J, Wei X-Y, Fan X (2017) Structural characterization of lignin and its degradation products with spectroscopic methods. J Spectrosc 1–15
26. Lu YC, Lu Y, Fan X (2020) Structure and characteristics of lignin. In: Sharma S, Kumar A (eds) Lignin. Springer Series on Polymer and Composite Materials, Springer, Cham, pp 17–75
27. Mansouri NE, Salvadó J (2007) Analytical methods for determining functional groups in various technical lignins. Ind Crops Prod 26(2):116–124
28. Nayak KK, Parkhey P, Sahu R (2020) Analysis of lignin using qualitative and quantitative methods. In: Sharma S, Kumar A (eds) Lignin. Springer series on polymer and composite materials. Springer, Cham
29. Nicholson DJ, Leavitt AT, Francis RC (2014) A three-stage Klason method for more accurate determinations of hardwood lignin content. Cellul Chem Technol 48(1–2):53–59
30. Pandey KK, Ramakantha V, Chauhan SS, Kumar ANA (2017) Current trends and future prospects in wood utilization, Wood is Good: Current Trends and Future Prospects in Wood Utilization, Springer, Singapore.
31. Pometto AL, Crawford DL (1988) High performance liquid chromatography of aromatic fragments from lignin degradation, Methods in Enzymology, vol 161 (no reference in text)
32. Popescu CM, Vasile C, Popescu MC, Singurel GH, Popa VI, Munteanu BS (2006) Analytical methods for lignin characterization. II. Spectroscopic studies. Cellul Chem Technol 40(8):597–622
33. Pralat K, Krymarys E (2017) A particle size distribution measurements of selected building material using laser diffraction method. Techn Transl 5:95–108

34. Prothmann J, Spegel P, Sandahl M, Turner C (2018) Indentification of lignin oligomers in kraft lignin using ultra-high performance liquid chromatography/high resolution multiple stage tandem mass spectrometry (UHPLC/HRMS^n). Anal Bioanal Chem 410:7803–7814
35. Rath S, Pradhan D, Du H, Mohapatra S, Thatoi H (2024) Transforming Lignin into value-added products: perspectives on lignin chemistry, lignin-based biocomposites, and pathways for augmenting lignolytic enzyme production. Adv Compos Hybrid Mater 7(27)
36. Research GV (2020) Lignin market size, share & trends analysis report by product (lignosulphonates, kraft, organosolv), by application (macromolecule, aromatic), by region, and segment forecasts 2020–2027. https://www.grandviewresearch.com/industry-analysis/lignin-market
37. Reyes-Rivera J, Terrazas T (2017) Lignin analysis by HPLC and FTIR. Methods Mol Biol 1544:193–211
38. Ruwoldt J (2022) Emulsion stabilization with lignosulphonates. In: Sand A, Tuteja J (eds) Lignin-chemistry, structure and application. https://doi.org/10.5772/intechopen.107336
39. Sharma S, Nadda AK (2020) Lignin: biosynthesis and transformation for industrial applications. Springer Nature, Technology & Engineering
40. Shorey R, Salaghi A, Fatehi P, Mekonnen TH (2024) Valorization of lignin for advanced material applications: a review. RSC Sustain 2(84):804–831
41. Stark NM, Yelle DJ, Agarwal UP (2016) Techniques for characterizing lignin. In: Lignin in polymer composites, pp 49–66
42. Sygula E, Ciolkosz D, Bialowiec A (2024) The significance of structural components of lignocellulosic biomass on volatile organic compounds presence on biochar—a review. Wood Sci Technol 58:859–886
43. Technical Association of the Pulp and Paper Industry (2002) Acid-insoluble lignin in wood and pulp (Official Test Method T 222 om-02). Atlanta, GA: TAPPI Press. https://www.taylorfrancis.com/books/mono/10.1201/EBK1574444865/lignin-lignans-cyril-heitner-dimmel-john-schmidt
44. Agarwal UP, Atalla RH (2010) Vibrational spectroscopy. In C. Heitner, D. R. Dimmel, & J. A. Schmidt (Eds.), Lignin and Lignans: Advances in Chemistry (pp. 103–136). Boca Raton, FL: CRC Press
45. Van den Oever M, Abaecherli, A (2023) Determination of particle size distribution for lignin powders. Method AnF005, International Lignin Institute, March 2023
46. Wang Q, Jahan MS, Liu SS, Miao QX, Ni YH (2014) Lignin removal enhancement from prehydrolysis liquor of kraft based dissolving pulp production by laccase induced polymerization. Bioresour Technol 164:380–385
47. Weng JL, Lun SL, Xue BL, Sun RC (2013) Recent advances in characterization of lignin polymer by solution-state nuclear magnetic resonance (NMR) methodology. Materials 6(1):359–391
48. Xia Z, Akim LG, Argyropoulos DS (2001) Quantitative (13)C NMR analysis of lignins with internal standards. J Agric Food Chem 49(8):3573–3578
49. Xu F, Jhong XC, Sun RC, Lu Q (2006) Anatomy, ultrastructure and lignin distribution in cell wall of *Caragana Korshinskii*. Ind Crops Prod 24(2):186–193
50. Yu H, Gui C, Ji Y, Li X, Rao F, Huan W, Li L (2022) Changes in chemical and thermal properties of bamboo after delignification treatment. Polymers 14(13):2573
51. Zhou N, Thilakarathna WPDW, He QS, Rupasinghe HPV (2022) A review: depolymerization of lignin to generate high value bio-products; opportunities, challenges and prospects. Front Energy Res 9:1–18

Modification of Black Liquor Properties to Enhance the Chemical Recovery Efficiency

A. K. Dixit and B. P. Thapliyal

Abstract The black liquors generated from different raw materials have different ratios of organic to inorganic constituents, varying concentrations of process and non-process elements, widely ranging viscosities, and thermal properties. These situations can result in inefficient operations of chemical recovery systems, and over a long period can lead to terrible situations with the build-up of many problems in almost all unit operations. Modification of black liquor properties by subjecting it to controlled chemical, mechanical, and thermal treatments is one of the solutions to make black liquors suitable for smooth operation in chemical recovery unit operations. The agro residue and bamboo based black liquors are generally difficult to handle. Bagasse, wheat straw, rice straw and bamboo black liquors have some unusual characteristics of high viscosity and scaling during evaporation in multiple effect evaporators. As a result, they pose problems in subsequent recovery operations like combustion in recovery boiler, clarification of white liquor and reburning of lime sludge in lime kiln. Firing of low solids black liquor due to highly viscous semi-concentrated black liquor, significantly affects the thermal efficiency of recovery boilers and results in uneconomical operation. Silica present in straw and bamboo black liquors causes problems in smooth evaporation in multiple effect evaporators (MEEs), due to poor heat transfer and equipment down time. Liquor heat treatment and desilication of black liquor are discussed in this chapter to address these problems by modifying black liquor properties after subjecting liquors to chemical and thermal treatments. The impact of modification of black liquor properties is evident on operation of various other chemical recovery processes.

Keywords Thermal efficiency · Multiple effect evaporator · Scaling · Liquor heat treatment · Lignin carbohydrate complexes · Silica · Desilication · Carbonation

A. K. Dixit
Central Pulp and Paper Research Institute, Saharanpur, Uttar Pradesh , India

B. P. Thapliyal (✉)
Indian Agro and Recycled Paper Mills Association, New Delhi, India
e-mail: bipin_thapliyal@yahoo.com

S. Sundaramurthy et al. (eds.), *Black Liquor Waste Management in the Pulp and Paper Industry*, Sustainable Materials and Technology,
https://doi.org/10.1007/978-981-96-4237-3_4

1 Introduction

Black liquor is a valuable by-product generated from pulping operation and its efficient utilization in the chemical recovery systems influences the sustainability of the entire paper mill. Therefore, while an appropriate understanding of black liquor characteristics is vital for chemical recovery operations, modification of its properties by chemical and thermal treatments plays an important role in efficient utilization in various chemical recovery unit operations. Black liquor has been subjected to various modifications to process it efficiently in the recovery systems. In this chapter, we will discuss some of the efforts made by our research group at Central Pulp and Paper Research Institute (CPPRI), Saharanpur, India, which is actively working on modification of black liquor properties to process it efficiently in recovery operations. Some of the black liquor modification studies conducted by our group to modify black liquor properties at a laboratory scale and demonstrated at a pilot scale are discussed in this chapter. The modification studies are related to thermal treatment and desilication of agro-residue black liquors.

2 Physico-chemical Characteristics of Black Liquor

Black liquor is the spent pulping liquor from the cooking operation of raw material. It is an aqueous solution of NaOH, Na_2S, and dissolved organics. It contains dissolved components of raw material along with spent pulping chemicals and water. The physical and chemical properties of black liquor have a significant effect on the performance of the chemical recovery system. The black liquor contains some residual active alkali from the cooking and normally has a pH of 12.0. If there is insufficient residual alkali in the black liquor, there is an increased possibility of lignin precipitation when high solid concentration is reached [1]. The variety of raw materials and diverse pulping operations result in a black liquor with wide range of chemical and physical characteristics.

The chemical composition of black liquor, particularly the extent and nature of lignin, its stability and hemicellulose content has pronounced influence on the chemical recovery operation and selective precipitation of silica [2, 3]. The chemical analysis of black liquor generated from different raw materials showed a wide variation in the chemical behavior [4]. In most of the small straw-based mills, black liquor has very little residual active alkali which results in sludge formation which can become a source of scale in evaporators. The straw black liquors exhibit a higher foaming tendency compared to wood due to a high proportion of organic acids [4]. Keeping in view of the wide variation in black liquor properties it is important to know these properties prior to optimizing the thermal treatment and desilication parameters for different black liquors.

The nature of black liquor generated depends largely on the pulping conditions employed and the quality of raw material [5]. Bamboo and straw pulping conditions have already been established and the mills based on bamboo and wheat straw raw materials practice chemical recovery, while in the case of rice straw, no mills have chemical recovery so far due to the presence of very high silica in black liquor. Therefore, pulping conditions must be optimized to have a stable black liquor and acceptable quality of pulp produced [6]. Due to economic reasons, rice straw is cooked with low alkali and the properties of black liquor generated are not acceptable from the recovery point of view. Black liquor should be stable if it must be processed in a chemical recovery plant for recovery of chemical and thermal energy. The stability of black liquor depends on certain physico-chemical and rheological properties of black liquor. The residual active alkali (RAA) of black liquor should be a minimum of 5–7 gpl for its stability and pH should be around 12.0 for the smooth operation of the chemical recovery and desilication plant. Rice straw-based mills do not have any chemical recovery plant (due to the presence of silica in black liquor); hence, pulping of rice straw is carried out keeping a low alkali charge (around 6–7% as NaOH) as it is not going to be recovered after pulping. The black liquor generated with a low alkali charge has a high pH value (9.5–10.0) and almost nil RAA. Pulping of rice straw was carried out with different chemical charges in order to have a black liquor of desired properties.

During the alkaline pulping process, cooking chemicals ($NaOH + Na_2S$) partly convert silica into soluble components (silicates), while the remaining stays with the pulp [7]. The amount of silica dissolved in black liquor depends on the amount of alkali charged and silica in the raw material. Thus, pulping conditions play an important role in the extent of dissolution of silica in black liquor and the characteristics of black liquor.

2.1 *Alkaline Pulping Processes*

Fibrous raw materials are composed of cellulose, lignin, and extractives. The main aim of pulping is to separate the cellulosic fibers from other components of raw material. Different kinds of pulping processes like chemical, semi-chemical, and mechanical pulping processes are employed depending upon the type and end-use of pulp to be produced. Only in the chemical pulping process maximum lignin and extractives are dissolved leaving behind cellulosic fibers. For writing, printing, and other grades of paper, chemical pulps are used. Different chemical methods like– the soda process involving cooking with sodium hydroxide, the sulfite process with calcium or sodium bisulfite, and the kraft process with a mixture of sodium hydroxide and sodium sulfide, are available. The Kraft process is a predominant one among all chemical pulping processes. The pulping treatment involves heating the fibrous raw material in an aqueous solution of sodium hydroxide and sodium sulfide at a cooking temperature of about 170 °C for a period of one to two hours. During this treatment, lignin is extensively degraded and the degradation products are solubilized. Carbohydrates, in particular, hemicelluloses, undergo partial degradation and dissolution, while the extractive fractions are largely removed.

The alkaline pulping process is a complex process affected by many physical and chemical factors such as the morphological characteristics of the raw material, the penetration pathway of the chemicals, the swelling or thermal coalescence of the lignin-containing interface, the cleavage of lignin-carbohydrate bonds, the degradation of the lignin macromolecule, the tendency of the fragments to enter secondary condensation reactions and simultaneous undesirable carbohydrate degradation reactions [8, 9].

2.2 Reactions in Alkaline Pulping

When looking into the anatomy of the fibrous raw materials, it is observed that cellulose fibers are cemented by polymeric lignin in the middle lamella portion. The basic principle of chemical pulping is to solubilize this lignin thereby separating cellulose fibers. This process is called delignification. Delignification involving lignin degradation is accomplished to a varying extent by the degradation of carbohydrate constituents i.e., cellulose and hemicelluloses.

2.3 Lignin-Nature and Structure

Cellulose fibers are bound by cementing material lignin with inclusions of hemicellulose, which differ in their chemical nature. Cellulose is a high molecular weight linear polymer. Hemicelluloses are low molecular weight polysaccharides and constitute the matrix substance. Lignin is a three-dimensional cross-linked polymer consisting of randomly polymerized p-hydroxyphenyl propane (Trans-p-coumaryl alcohol), guaiacyl propane (Trans-p-coniferyl-alcohol), and syringyl propane (Trans-p-sinapyl alcohol) units. It occurs in the cell walls of true vascular plants, ferns, and club mosses but not in the mosses, algae, and microorganisms [10]. Lignin is distributed as a matrix component with hemicelluloses, probably linked with a covalent bond [11] in the space of inter-cellulosic microfibrils in primary and secondary cell walls and the middle lamella of fibers [12, 13]. It acts as an encrusting and cementing material of fibers, imparts hardening and mechanical strength to the xylan tissue [14], regulates water and nutrient transportation in plants, and protects them from micro-organisms. In nature, it is abundant, in terms of weight probably second to cellulose among the renewable organic materials. In terms of energy content, it might be the single most abundant [15].

The main precursor of softwoods (gymnosperms) lignin is guaiacyl propane (Trans-p-coniferyl-alcohol), while in the hardwoods (angiosperms), lignin is composed of guaiacyl propane (Trans-p-coniferyl-alcohol) and syringyl propane (Trans-p-sinapyl alcohol) units. Grasses and straw lignin are composed of p-hydroxyphenyl propane (Trans-p-coumaryl alcohol) in addition to the above [16,

17] and p-coumaric acids with the esterified terminal hydroxy group of the side chain [18].

2.4 Lignin Reactions in Pulping

The elementary building blocks of all lignins are phenyl propane units interconnected by C-O and/or C–C bonds [19]. Reactions of lignin during kraft pulping can be divided into two categories-first, the degradation reactions which lead to the liberation of lignin fragments and ultimately to their dissolution, and the second, condensation reactions which increase the molecular size of the lignin fragments and may result in their precipitation [20]. Extensive studies, during the past few decades, have been carried out using model compounds to elucidate the mechanism of these reactions [21, 22]. During pulping, most of the carbon–carbon bonds are stable whereas carbon–oxygen bonds are cleaved to varying extents. Two main types of ether linkages in lignin phenolic aryl propane units exist, α-aryl ether bonds and β-aryl ether bonds. Due to the high frequencies of α-aryl ether linkages, the alkali-promoted and sulfidolytic cleavage reaction plays a dominating role in the fragmentation of lignin during the kraft pulping process. Only minor contributions are from β-aryl ether linkages in the non-phenolic units [23].

2.4.1 Degradation Reactions

α-aryl ether bonds are readily cleaved by the reactions involving an alkali-promoted rearrangement of phenolate structure into corresponding quinone methide structure with the elimination of α-aroxy substituent [19, 24]. The rate is independent of the hydroxide ion concentration provided the phenolic hydroxyl group in the substrate is completely ionized [25]. Phenolic units containing β-aryl ether bonds undergo some initial transformation into intermediates of quinone-methide type by alkali-promoted elimination α–hydroxy or aroxyl groups. In kraft pulping, the presence of hydrosulfide and sulfide ions (SH-, S-) as reactive species in sulfate cooking liquor has to be taken into account. These ions are weaker but stronger nucleophiles than hydroxyl ions. So quinone methide formation is followed by the reversible addition of hydrosulfide ions giving benzyl mercaptide structure. The mercaptide anion then intra-molecularly attacks the β-carbon atom forming a thiirane intermediate with the elimination of the β–aroxy substituent (β-aryl ether cleavage) [26]. The thiirane structure eliminates elemental sulfur affording the corresponding p-hydroxy styrene structure. Delignification is faster in the kraft process compared to the soda process due to the fact that intermediates react with stronger nucleophiles (SH-, S-) rather than stronger bases (OH-).

Thus, the main reactions responsible for lignin fragmentation are the cleavage of the α-aryl ether bonds in phenolic units via quinone methide intermediate and cleavage of β-aryl ether bonds in phenolic units via quinone methide and episulphide intermediates. When all the phenolic units of the α and β-aryl ether types have reacted further, lignin degradation requires drastic conditions in which β–aryl ether linkages in non-phenolic units are cleaved in the bulk phase of delignification [27]. In the last phase, i.e. residual phase, possible cleavage of carbon–carbon bonds takes place via alkali-promoted retrograde aldol-type reactions. Thus, the initial phase comprises the pulping treatment up to a temperature of about 150 °C and results in the dissolution of 20–25% of the total amount of lignin content. The bulk phase at 150–170 °C results in the dissolution of about 60% lignin and finally the residual phase at 170 °C results in the dissolution of 10–15% lignin.

The cleavage of β-aryl linkages in the phenolic units is comparatively more resistant towards alkali degradation and cleavage of these bonds takes place via oxirane intermediate at high temperature and hydroxide ion concentration with the elimination of β-aroxy substituent [24].

Besides, sulphidolytic cleavage of β –aryl ether linkages in the nonphenolic units containing an α-carbonyl group [28], alkali-promoted cleavage of C–C bond via retrograde aldol reactions [29] and cleavage of methyl-aryl ether linkages by hydrosulphide ion [30] leading to the formation of an additional phenolic hydroxyl group in lignin molecule, also plays role in the delignification process. Besides, above main delignification reactions, there are many more minor degradation reactions governing the delignification.

Two black liquor modification studies tried by our research group have been successful and demonstrated on a pilot scale in the mills. The studies are liquor heat treatment (LHT) and desilication of black liquor which will be discussed in this chapter. The objective of liquor heat treatment is to reduce the viscosity of black liquor to achieve higher solids of low-viscosity flowable black liquors after evaporation and prior to incineration in the recovery boiler. The advantages of LHT are higher steam generation per ton of black liquor solids and reduced environmental impact because of higher solids incineration in the recovery boiler. LHT is beneficial for bagasse black liquor, however, it has significant advantages in hardwood black liquor. The desilication technology is also helpful in the efficient operation of evaporators, recovery boilers, causticization, and lime mud re-burning, along with the recovery of valuable high-grade amorphous silica as a by-product. The technology was initially tested on bamboo black liquors and later on wheat and rice straw black liquors, having higher silica contents. The results on modification of black liquor chemical, physical, and combustion properties after LHT and desilication treatments with advantages and disadvantages are presented in the next sections of the chapter.

3 Thermal Treatment for Viscosity Reduction

The black liquor viscosity characteristics, specifically from non-wood-based pulp and paper mills is such that, even at normal operating temperatures of 100–115 °C, the viscosity at 55% *w/w* total solids concentrations exceeds 500 cps. Thus, the higher viscosity problems lead to the firing of black liquor at low solids concentration and thereby lower steam generation.

One of the reasons for the high viscosity in straw black liquors is the higher silica content. Studies have shown that in the case of bagasse black liquor, high molecular weight organic residues like polysaccharides are linked with lignin-carbohydrates complexes (LCC's) and the high molar mass lignin molecules are mainly responsible for abnormally high viscosities of bagasse black liquors.

Thermal treatment of bagasse and straw black liquor was found to reduce the viscosity substantially. Major benefits of liquor thermal treatment include more than 60% reduction in viscosity, better combustion of black liquor, higher steam generation, and reduced emissions. In this section, the modification of bagasse and straw black liquor viscosity by thermal treatment is discussed along with its impact on various other chemical and combustion properties.

It is well known that the firing of black liquor at high dry solids concentration with proper combustion is one of the prerequisites for the generation of high-pressure steam and efficient cogeneration. Modification of black liquor properties through thermal treatment has been the priority area for the smooth and effective processing of black liquor to improve the energy efficiency of the chemical recovery systems. In India wheat and bagasse-based mills are operational and to combat the problems related to their low efficient chemical recovery operation. Besides, studies on thermal treatment of non-wood black liquor have been successfully developed and demonstrated by Indian research groups at CPPRI, both on laboratory and pilot plant scale.

The results obtained during pilot plant trials were encouraging in respect of viscosity and combustion behavior. The viscosity was reduced to more than 60% (i.e., to 398 cps as against 1000 cps at a solid concentration of 65% *w/w* at 98 °C temp.). The combustion behavior of black liquor was also improved as indicated by the improved swelling behavior of black liquor since the swelling volume ratio was increased from 8 ml/g to 17 ml/g. The results of pilot plant trials have given confidence in up-scaling the technology to commercial-scale operation with an option of retrofitting in existing and/or upcoming chemical recovery installations in the non-wood-based pulp and paper mills.

In most of the non-wood mills, black liquor is evaporated to a concentration level of around 58–63% *w/w* followed by its combustion in a chemical recovery furnace as against 75–80%. Chemical recovery, which is a complex operation, has several dependent and independent variables. Among these, the black liquor concentration and the combustion air are two important variables, that have an influence on the thermal efficiency of the chemical recovery boiler and the opportunities for co-generation. The generation of high-pressure steam is one of the prerequisites for co-generation. However, one of the present limitations in having high-pressure–temperature steam in Indian paper mills has been, the firing of the black liquor at lower solids concentration, mainly due to higher viscosities of the black liquor resulting from non-wood-based mills. Thus, for effective steam generation and for improved co-generation, a need to adopt high-solids black liquor combustion was felt. Therefore, thermal treatment of black liquor process was successfully developed and demonstrated on a pilot scale for non-wood-based black liquors which could help in a drastic reduction of black liquor viscosity thus, enabling to fire the black liquor at high dry solids concentration. The next sections highlight the results of the pilot plant trials conducted on the thermal treatment of black liquor from an integrated pulp and paper mill employing bagasse and Eucalyptus as major fibrous raw materials.

3.1 Black Liquor Viscosity

The black liquor viscosity characteristics, specifically from non-wood-based pulp and paper mills are such that even at a normal operating temperature of 100–115 °C, the viscosity at 55% *w/w* total dissolved solids concentration exceeds 500 CPS [31]. Figure 1 shows the comparative viscosity results of various wood and non-wood-based black liquors which show that non-wood black liquor including bagasse and bamboo have very high viscosities as compared to softwood and hardwood black liquors. These higher viscosities in the case of non-wood-based black liquors make it impractical to handle such thick black liquors. Thus, the higher viscosity problems lead to the firing of black liquor at low solids concentration and thereby lower steam generation.

With increased understanding and in-depth studies on basic aspects of black liquor, it could be inferred that high molecular weight organic residues like polysaccharides

Fig. 1 Viscosity curves of black liquors at 90 °C

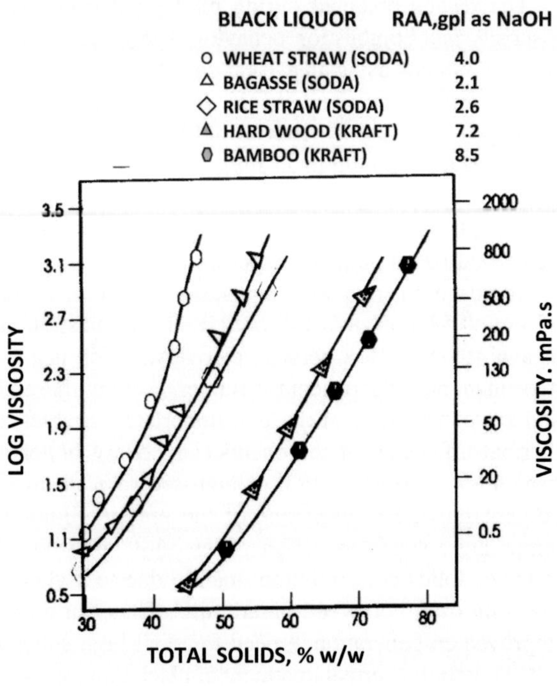

Table 1 Effect of lignin carbohydrate complexes (LCC'S) on viscosity of bagasse black liquors [32]

Particulars	Black liquor, before LCC separation	Black liquor, after LCC separation
Total solids, % *w/w*	Viscosity, mPa.s at 90 °C	
55	252	21
60	478	50
65	1000	320
68	Could not be measured	410

linked with lignin molecules, i.e., Lignin–Carbohydrate Complexes (LCC's) and the high molar mass lignin are mainly responsible for abnormally high viscosities of black liquors [32]. This is supported from the determination of the viscosities of the black liquor after the removal of the LCC'S from the untreated and treated black liquors. The results are shown in Table 1. From the results, it is clearly seen that the separation of LCC'S from the black liquors helps in a drastic reduction of viscosity.

3.2 *Thermal Treatment of Black Liquor*

Increased understanding of the basic black liquor properties of wood and non-wood raw materials during the last 20 years has led to the development of the process for thermal treatment of black liquor. The process basically incorporates heat treatment of semi-concentrated black liquor at temperatures between 180 and 190 °C in the presence of a certain amount of residual active alkali for the specified time period, which results in depolymerization of high molecular weight lignin carbohydrate complexes into lower mass components. The findings are supported by gel chromatographic studies as shown in Fig. 2. From the chromatogram, it is clear that the thermal treatment of the black liquor resulted in the cleavage of Lignin-Carbohydrate Complexes which caused a decrease in the molar mass of the lignin entity which appears in the form of a new peak of relatively lower molar mass fraction, eluting out at elution volumes of 60 and 68 ml as compared to the original, where only a single peak is observed.

As a result of the thermal treatment of black liquor, there was a remarkable reduction in black liquor viscosity making it possible to achieve higher dry solids (more than 70% *w/w*) in the remaining bodies of the multiple effect evaporator (MEE) itself and eliminates the need of direct contact evaporator. Reduction in viscosity after thermal treatment could help in handling the black liquor by keeping it in a less viscous form even at atmospheric pressure.

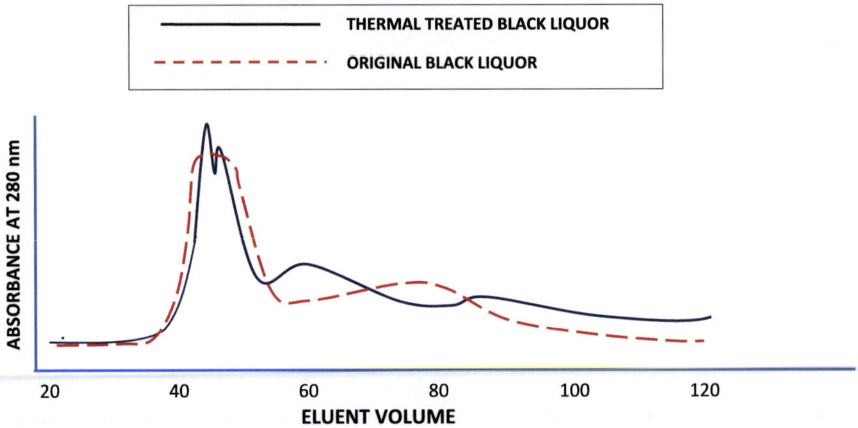

Fig. 2 Gel chromatogram showing influence of thermal treatment on soda bagasse black liquor (by using Sephadex G25 column)

3.3 Pilot Plant Trials

Based on data generated from laboratory studies, a pilot plant was designed for processing 200 l/h of black liquor. The process flow sheet of the pilot plant is shown in Figs. 3 and 4. The pilot plant consists of the basic components as shown in Table 2. The pilot plant was run with black liquor procured from an integrated bagasse and eucalyptus-based mill in order to test the response of thermal treatment on the mill's black liquor and to generate data for exploring the possibilities of retrofitting the thermal treatment process in existing chemical recovery system.

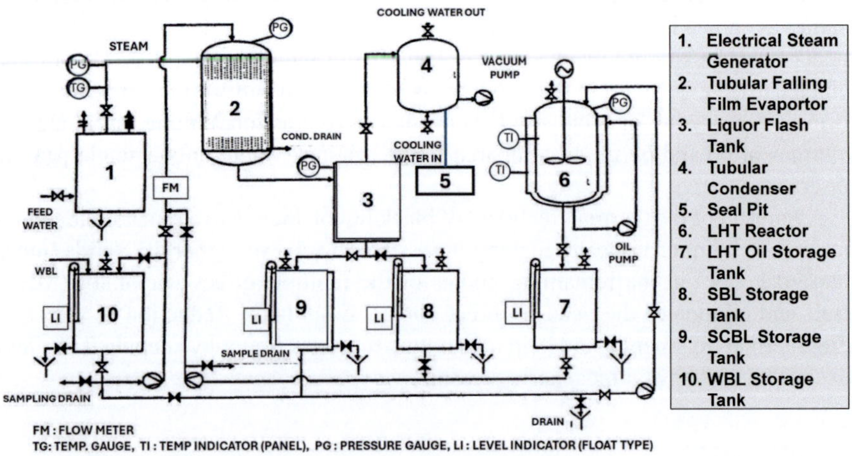

Fig. 3 Schematic flow sheet of black liquor thermal treatment plant

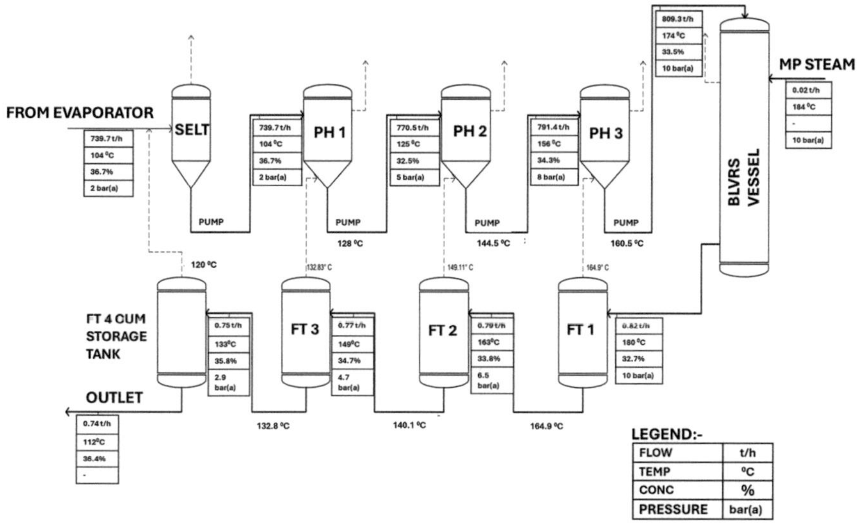

Fig. 4 Material and energy balance process flow sheet of black liquor thermal treatment plant

Table 2 Various units of pilot plant for thermal treatment of black liquor

S. No.	Description of the components	Capacity
1	S.S. Reactor vessel with MS heating Jacket having oil recirculation to raise temperature of black liquor from 200 to 220 °C. Equipped with agitator at the top, gear box and motor and appropriate pressure and temperature gauges	40 l
2	Falling film evaporator, 2 m height having 10 tubes of 1″ OD for working with 5 kg/cm^2 steam pressure	Heating surface area–1.2 m^2
3	Storage tank made of MS, 4 Nos	200, 100, 100 and 80 l
4	Shell and tube M.S condenser for condensing vapor coming out from Calendria	–
5	S.S vapor separator	50 kg
6	S.S recycling gear and centrifugal pumps for transfer of weak and thick black liquor	3 Nos
7	Vacuum pump, 5 HP	1 No
8	Gas holder for non-condensable gases	1 No

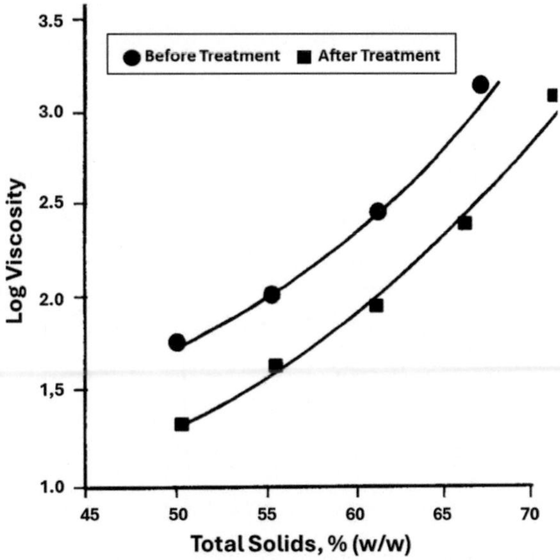

Fig. 5 Viscosity results of pilot plant trials during thermal treatment of mill's black liquor (bagasse 85% + eucalyptus 15%)

3.3.1 Effect of Thermal Treatment on the Viscosity and Combustion of Black Liquor

The results of the pilot plant trial for thermal treatment of black liquor for viscosity reduction and combustion behavior are shown in Fig. 5 and Table 3 respectively. From the results of viscosity, it is evident that there is a drastic reduction in viscosity of the black liquor after thermal treatment which was reduced by more than 50%. Reduction in viscosity could make it possible to evaporate the black liquor in the evaporator to a solids concentration level of more than 70% (*w/w*) while maintaining the viscosity level of 350 mPa.s. This reduction in viscosity could help keep the black liquor in a form, that can be easily pumped even at higher solids concentration and at atmospheric pressure allowing storage of the thick black liquor in atmospheric tanks.

Reduction in viscosity further helps in reducing the requirement of electrical energy during the pumping of thick black liquor. It is evident from the fact that a 30% increase in bagasse black liquor viscosity results due to a remarkable increase in the requirement of pumping energy, almost to the tune of 200% higher than softwood

Table 3 Results of combustibility before and after thermal treatment of black liquor

Particulars	Black liquor before thermal treatment	Black liquor after thermal treatment
Calorific value, KJ/kg	14,200	13,990
Swelling volume, ml/g	8.0	17.0
Temp. of ignition, °C	765	735

or hardwood black liquor in the same solids concentration. When compared to the softwood and hardwood black liquors, the comparative pumping energy requirement is 33% for softwood black liquor, whereas it is 63% for Indian hardwood black liquors.

3.3.2 Effect of Time on Viscosity Reduction

Liquor heat treatment is a thermal treatment method whereby residual alkali reacts with dissolved polysaccharides and lignin, and hemicellulose molecules at high temperatures (180–190 °C) for a certain period of time. This leads to the destruction or depolymerization of the high molecular weight compounds and hence a reduced viscosity. This viscosity reduction is permanent since the high molecular weight lignin and polysaccharide contents of the liquor have significant impacts on viscosity. In this process the intermediate liquor from second effect evaporator outlet at 36% solid content is heated to 180–185 °C and kept for 20–30 min retention time. The heavy molecular weight organic compounds de-polymerize and due to which the viscosity of black liquor gets reduced. The viscosity reduction is proportional to the degree of de-polymerization. Silica present in the black liquor retards the process of de-polymerization and to enhance the de-polymerization, addition of caustic is practiced.

Time for thermal treatment of black liquor was found to have a pronounced effect on the extent of reduction in viscosity, as evident from the results shown in Fig. 6. It is clearly seen that increasing the time of thermal treatment of black liquor from 20 to 35 min, there is improvement in reduction in viscosity however, beyond 35 min, the black liquor becomes unstable resulting in phase separation and making it difficult to evaporate further.

The results of thermal treatment trials on viscosity of bagasse black liquors at 175, 180 and 185 °C are presented in Table 4. In these trials the bagasse black liquors at different solids level, ranging from 40 to 60%, were subjected to thermal treatment at different temperatures from 15 to 25 min. Table 5 shows that more than 50% reduction in viscosity of bagasse black liquor is achievable by treating upto 25 min at these temperatures.

3.3.3 Effect of Treatment on the Combustibility of Black Liquor

The combustibility of the black liquor is highly influenced by the thermo-plasticity of the organic residues in black liquor which is related to molecular size and uniform dispersion of the molecules. This influences the swelling characteristics of the black liquor during combustion. Swelling of the black liquor droplet during firing into the recovery furnace determines how the black liquor will behave during combustion. Higher swelling is an indication of better combustibility.

Results in Table 3 show the swelling of the thermally treated black liquor was improved from 8 ml/g to 17 ml/g which is an indication of improved combustibility

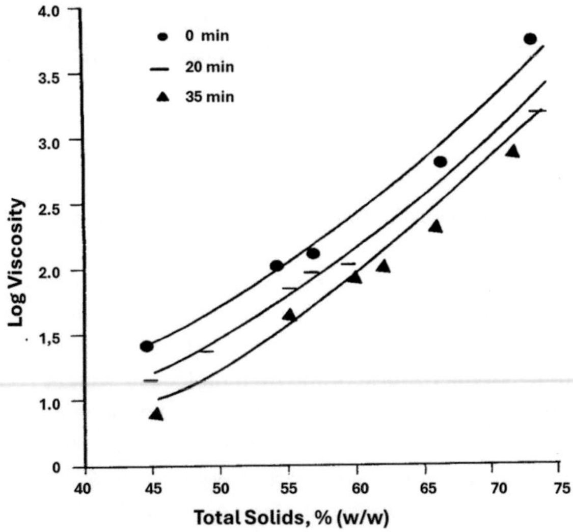

Fig. 6 Effect of time on reduction of viscosity during thermal treatment of mill's black liquor (bagasse 85% + eucalyptus 15%)

Table 4 Effect of temperature and time on viscosity of control and treated black liquors at different total solids

Total solids, %w/w	Viscosity (cP) at 100 °C									
	Control black liquor viscosity, (cP)	Treated								
		175 °C			180 °C			185 °C		
		15 min	20 min	25 min	15 min	20 min	25 min	15 min	20 min	25 min
40	27.82	13.36	13.23	13.25	12.95	12.74	12.75	12.83	12.63	12.55
45	65.08	30.70	30.26	30.49	29.78	29.19	29.19	29.49	28.68	28.65
50	138.8	61.00	59.56	59.69	59.12	57.88	57.76	58.45	57.17	56.94
55	368.1	159.82	156.07	156.03	154.53	151.03	151.29	153.13	167.34	168.74
60	986	434.92	423.88	423.59	421.22	417.28	415.30	419.64	413.04	414.22
60.05	1002	446.79	431.96	435.37	432.46	425.45	424.35	431.06	421.94	424.85

after thermal treatment of black liquor. The results of thermal treatment pilot plant trials on bagasse black liquor's swelling volume ratio after treatment for different temperatures and time are presented in Fig. 7.

The differential thermal analysis of the two black liquors also shows a reduction in the Temperature of Ignition (Tig.) in thermally treated black liquor. It was found that it reduced from 765 to 735 °C. The results clearly indicated that the combustibility of the black liquor was improved in terms of swelling behavior and temperature of ignition after thermal treatment.

Table 5 Effect of temperature and time on % reduction of viscosity of control and treated black liquors at different total solids

Total solids, %w/w	% reduction in viscosity									
	Control black liquor viscosity, (cP)	Treated								
		175 °C			180 °C			185 °C		
		15 min	20 min	25 min	15 min	20 min	25 min	15 min	20 min	25 min
40	27.82	51.98	52.46	52.39	53.44	54.21	54.18	53.90	54.60	54.88
45	65.08	52.82	53.51	53.15	54.24	55.14	55.15	54.68	55.93	55.97
50	138.8	56.05	57.09	56.99	57.41	58.30	58.39	57.90	58.81	58.98
55	368.1	56.58	57.60	57.61	58.02	58.97	58.90	58.40	54.54	54.16
60	986	55.89	57.01	57.04	57.28	57.68	57.88	57.44	58.11	57.99
60.05	1002	55.41	56.89	56.55	56.84	57.54	57.65	56.98	57.89	57.60

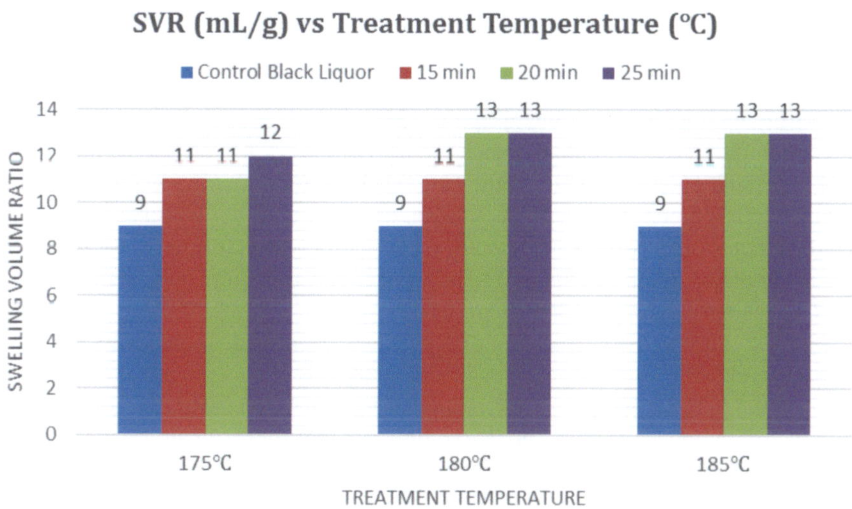

Fig. 7 Impact of thermal treatment temperature and time on swelling volume ratio of the bagasse black liquors

4 Desilication of Silica-Rich Black Liquors

Worldwide pulp and paper industry utilizes wood as the most widely used raw material for pulp and paper production. The availability of wood for pulp and paper production is limited in many countries due to excessive deforestation. Therefore to preserve the depleting forests, the paper industry in Asia and other parts of the world utilizes non-woody fibrous raw materials. China and India are the leading countries using a substantial proportion of non-wood raw materials such as bamboo, cereal straws, bagasse, etc., for pulp and paper production. In India, the share of agricultural residues is about 10% of total production. The non-wood agro residue-based raw materials are characterized by high silica content and it varies between 1.5 and 15%. The silica is dissolved in spent pulping liquor during the alkaline pulp-making process and the presence of this silica in black liquor creates innumerable problems in operating the chemical recovery systems. Evaporator, Recovery furnace, causticization, and lime kiln operations are adversely affected by the presence of silica in black liquor. The presence of silica in lime sludge adversely affects the lime mud reburning and in a wheat straw-based mill it is difficult to install a lime kiln on account of high silica contents in black liquor. The major reason is that the presence of silica in the re-burnt lime (CaO) is not suitable for recausticization and has to be discarded as solid waste from the pulp mill. Studies have been carried out for the removal of silica from pulping and chemical recovery loops for the last five decades as the presence of a higher quantity of silica, a non-process element, has been a matter of great concern for processing the black liquors in chemical recovery unit operations. To overcome the problems, desilication of black liquor by selective precipitation of silica is one of the preferred techniques.

Various methods of desilication have been reported in the literature by using chemicals and mineral acids to reduce the pH of black liquors and precipitate silica selectively without precipitation of lignins [33].

Carbon dioxide absorption by black liquors [34, 35] is also one of the options to reduce pH and precipitate silica. Lowering of pH by CO_2 absorption has also been used for lignin precipitation, as the removal of lignin makes it possible to increase the production capacity of the pulp mill, without increasing the load in the recovery boiler. Extracted lignin can be utilized as a viable substitute for solid fuels or can be used as a feedstock for producing various chemicals.

Initial studies on the desilication of high silica kraft black liquors were reported by Isono and Ono [36, 37] by carbon dioxide treatment. They reported a relationship between the pH and the silica removal percentage by carbon dioxide absorption in kraft black liquors. Aujla et al. [38] have proposed a mechanism of silica precipitation by lowering pH in chemi-thermo mechanical pulping black liquors.

Research work has been carried out on the desilication of black liquor and green liquor using different techniques like the addition of cations (lime and alumina) and reduction of pH [38]. The methods developed by various researchers could not live up to the expectations and none of the techniques was scaled up to a commercial scale. The methods involving the addition of cations did not become popular as enormous quantities of sludge are produced when desilication is carried out by the addition of lime or alumina. This resulted in an additional problem of suspended residual flocculant (lime or alumina) which was not settled easily and could cause problems like the problem due to the presence of silica.

During the initial trials on the reduction of pH either by acid or CO_2 absorption, attempts were not successful due to some fundamental problems. The research group at CPPRI, Saharanpur observed that it is impossible to control the pH by mineral acids, therefore slow lowering of pH with the absorption of CO_2 was tried and found to be successful with some corrections in the process controls. Since the critical pH for silica precipitation is very close to the pH for lignin precipitation, therefore, conventional methods of gas–liquid mass transfer like the direct bubbling of CO_2, or the use of a counter-current packed bed column were not successful due to uncontrolled carbonation and foam flooding in the column [39, 40].

The research group at CPPRI conducted detailed studies on the desilication of non-wood black liquors to effectively remove silica. The method to remove silica is based on carbonation under controlled conditions using recovery boiler flue gas as a source of carbon dioxide. Pioneering work has been done by Kulkarni and his associates since 1988 in this field resulting in successful lab-scale pilot plants, and industrial-scale demonstrations [31–34, 40]. Controlled pH reduction leading to selective precipitation of silica without co-precipitation of lignin has been perfected in the Desilication Technology developed by this group. They have reported nearly 90% desilication and evaluated black liquor properties before and after desilication. They have reported various incentives for silica removal from the black liquor, including reduced scaling in evaporators leading to lower downtime and resulting in improved thermal properties of the black liquor to facilitate its burning in a recovery furnace. With enhanced efficiency of the recovery boiler, they have reported the generation of additional steam and power. Higher efficiency of boilers is possible due to improvement in rheological properties of black liquors, which helps in achieving higher dry solids after evaporation in multiple effect evaporators. In the causticization section benefits reported include better fluidity of smelt and fast settling of mud in white liquor clarifiers for producing clear white liquor without suspended lime particles. Improved calcination of lime sludge and lower fuel requirement in the lime kiln is also the additional benefit of silica removal from black liquor. Most of the studies are focused on wheat straw and bagasse, the two main agro-residue raw materials used in the Indian subcontinent. Silica content in bagasse black liquor was reported in the range of 0.4–0.5% while in wheat straw it was 4.0–5.0% *w/w*. Due to the presence of this high amount of silica processing of wheat straw black liquor in chemical recovery has been very difficult.

4.1 Desilication Technology

CPPRI initiated work on desilication in the early eighties and carried out extensive basic research on the mechanism of silica precipitation and reactions involved in desilication by carbonation. CPPRI adopted the carbonation technique essentially because this process does not generate sludge like the lime treatment method where massive quantities of sludge are generated.

Bamboo was one of the major raw materials in the period 1980–2000 and issues related to silica were faced by the industry. Bamboo is now used in very small quantities by the mills and at present bagasse and wheat straw are the major non-wood raw materials used by the Indian paper industry. Bagasse contains low amounts of silica, in the range of 0.4–0.5%, and wheat straw contains 4.0–4.5% silica. The pulping process used by most of the agro residue-based mills is the soda process. Looking into the problem faced by the wheat straw-based mills there was a need to prove the efficacy of the technology on wheat straw black liquor. CPPRI initiated work on the desilication of black liquor and has developed a desilication technology based on a stepwise carbonation technique. The performance of a full-scale desilication plant has been successfully demonstrated on bamboo black liquor. Studies were also carried out on wheat straw black liquor.

Black liquor desilication studies have been conducted on a pilot scale also. Figure 8a shows the pilot plant facility used for the desilication trials at CPPRI. The reactor consists of two communicating tubes about 2.5 m in height and 10 cm in diameter, through which a high-speed centrifugal pump circulates the black liquor to be treated. Near the upper end of the tube, provision is made for the upward-flowing liquor to be sucked in by the venturi action of the flue gas through two openings. The inlet flue gas flow is broken down into discreet bubbles, and the size and speed of these gas bubbles can be regulated with the help of control systems. The shearing action of the liquor absorbs the gas bubbles continuously by liquid–gas mass transfer and it is replaced by new bubbles through exposure of new surfaces for efficient mass transfer over the liquid/gas boundary.

For the laboratory studies, synthetic flue gas was prepared by mixing air and CO_2 in a gas blender shown in Fig. 8b. CO_2 concentration in flue gas was maintained around 10–12% and measured continuously by TESTO gas analyzer. The CO_2 concentration in recovery boiler flue gas is in the range of 8–12%. To simulate the semi pilot plant scale trials with recovery boiler flue gas obtained in the mill, synthetic flue gas having similar concentration of CO_2 was used. The pH of black liquor was measured

a b

Fig. 8 **a** Photograph of the semi pilot scale submerse bubble reactor. **b** Photograph of the gas blender for synthetic flue gas preparation at laboratory

regularly, and samples were drawn at different pH levels for analysis to determine the percent silica removal and residual active alkali content at different pH levels.

The pH of the circulating liquor can be continuously measured and recorded by temperature-compensated electrodes. Rotameters are used to determine and regulate the gas intake into each reactor. The reaction temperature can be maintained above 75 °C initially and later on maintained around 70 °C. Filtration of the silica particles precipitated in this dynamic mixing system can be carried out offline by using a laboratory vacuum system using ordinary filter papers. The plant can also be used in a continuous operation and about 60 L of black liquor can be carbonated in each trial.

Further with CPPRI's basic knowledge of non-wood black liquor characteristics, it became possible to evolve a technology for selective precipitation of silica without co-precipitation of lignin by adapting step-wise, gentle carbonation technique. The concept of a submerse Bubble reactor was developed by CPPRI for the desilication of black liquor. In this reactor the liquor is treated with gas in a tubular system by a pump, which on its downward path sucks the gas in the form of discreet bubbles. The shearing action of the flow works the gas bubble in a kneading fashion continuously, exposing new gas–liquid interfaces. A mohno pump was used for circulation of the black liquor and carbonated black liquor was collected from the bottom of the column.

The method was successfully tested on the semi-pilot scale for more than three years and subsequently scaled up by more than 1000 times (30 l/h to 3 40 m^3/h) and a commercial prototype mill size desilication plant in a bamboo-based mill was operated in 1989. Based on the technology another commercial-scale plant with a capacity of 25 m^3/h was commissioned in a rice straw-based mill in 2006. Silica content in original black liquor was in the range of 6–7% whereas desilicated black liquor contained silica in the range of 0.5–0.6% indicating more than 90% desilication efficiency.

4.1.1 Process Details

Carbonation of black liquor is the major reaction and controlled carbonation results in the efficient desilication process. The reactions that take place during carbonation are shown below.

Black liquor contains NaOH, Na_2CO_3, Na_2S, organically bound sodium (Na-Org,), Na_2SiO_3 and hemicelluloses. These components undergo following changes in aqueous medium during carbonation.

Alkali present as NaOH and Na_2S, undergoes following changes during the carbonation reaction.

i. $NaOH \rightarrow Na_2CO_3$
   $Na_2CO_3 \rightarrow NaHCO_3$
ii. $Na_2S \rightarrow NaSH$
   $NaSH \rightarrow H_2S$

Organic sodium present in black liquor contains Org.–O-Na and Org.-O-COONa. These organically bound sodium undergo following reactions on carbonation.

iii. Org.-O-Na → Org.-OH
Org.-O-COONa → Org.-COOH

Sodium silicate present in the black liquor also undergoes following change.

iv. $Na_2SiO_3 \rightarrow H_2SiO_3 + H_2SiO_4$

Further carbonation converts the silicates into silicic acid gels, as given below.

$$H_2SiO_3 + H_2SiO_4 \rightarrow (H_2SiO_3)_x + (H_2SiO_4)_x \text{(Silicic acid gels)}$$

Carbonation influences pH by lowering and it results in co-precipitation of lignin with silica. Lignin constitutes almost 50% of the total organic content in black liquor. The molecular size of alkali lignin increases due to condensation reactions during pulping of raw materials and it also affects the precipitation of lignin. Most important hydrophilic groups in the alkali lignin are –OH and COOH groups. The colloidal particles lack their stability when pH and RAA is reduced to such an extent that marked proportion of the ionized –ONa groups transform into-OH groups. Precipitation of these lignin groups takes place at following pH. Thus lowering the pH leads to precipitation of lignin. The critical pH range for silica precipitation is very close to the lignin precipitation pH range.

Group	**pH (25 °C)**
Phenolic-OH	9.4–10.8
Phenolic-COOH	about 5.0

Tables 6 and 7 present the results of studies on pH lowering in bamboo and rice straw black liquors to evaluate the co-precipitation of lignin with silica. The results show the onset of precipitation by lowering of pH in bamboo and rice straw black liquors.

In order to avoid the co-precipitation of lignin with silica precipitation, by lowering of black liquor pH, utmost care is taken to lower the pH during carbonation. The

Table 6 Effect of pH lowering on lignin co-precipitation in bamboo black liquor

Particulars	pH	Lignin content in silica sludge, %	Residual silica, g/l	% Desilication
Original black liquor	12.8	–	4.2	–
Carbonated black liquor	10.3	0.1	1.2	71.4
	10.2	0.3	0.8	80.9
	10.1	0.4	0.5	88.1
	9.9	1.0	0.4	90.5
	9.8	2.8	0.4	90.5

Table 7 Effect of pH lowering on lignin co-precipitation in rice straw black liquor

Particulars	PH	Lignin content in silica sludge, %	Residual silica, g/l	% Desilication
Original black liquor	12.0	–	13.3	
Carbonated black liquor	10.1	0.1	0.7	94.7
	10.0	0.2	0.4	97.0
	9.9	0.3	0.35	97.4
	9.8	0.7	0.3	97.8
	9.6	1.4	0.3	97.8
	9.5	3.5	0.26	98.0

desilication process, therefore, involves stepwise carbonation in three reactors for the reduction of pH to the desired level. In reactor 1 (SRT-1) pH is dropped from 12.5 to 11.5 and in reactor 2 (SRT-2) it is further dropped to 10.7 and finally, in reactor 3 (SRT-3) the pH is brought down to the desired level as optimized pH. Carbonated black liquor is taken in a clarifier for settling of precipitated silica. Underflow of the clarifier is taken to filter press whereas overflow mixed with press filtrate is taken into re-alkalization tank to raise the pH by addition of alkali. This re-alkalized black liquor is sent for evaporation in the MEE street. A schematic diagram of the above desilication technology is shown in Fig. 9.

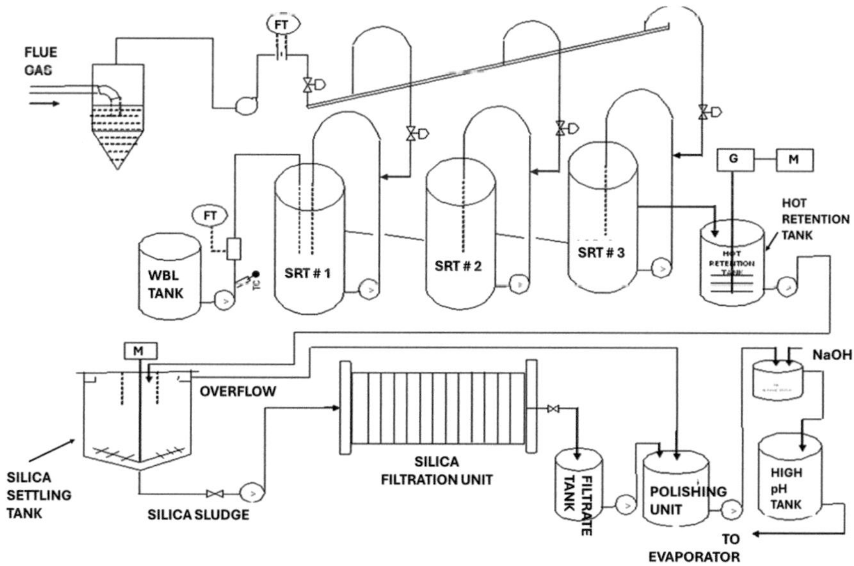

Fig. 9 Schematic of the desilication technology pilot plant

4.1.2 Mechanism of Silica Precipitation by Lowering of the pH

Silica is present in black liquor in the form of Na_2SiO_3 [37]. Lowering the pH of black liquor converts silicates into insoluble polymeric anhydrides of silicic acid. The solution in equilibrium with amorphous silica at ordinary temperatures contains monomeric monosilicic acid, $Si(OH)_4$. Lowering the pH of alkali silicate solutions results first in the formation of monosilicic acid [38]. This monomer undergoes polymerization, at a concentration greater than the solubility of amorphous silica (100–200 ppm), forming higher molecular weight species. The mechanism of polymerization is ionic and is proportional to OH concentration, above pH 2.0. Above pH 7.0, stabilized particles (Sols) grow to a size of about 100 nm. On the contrary, when salts are present, to neutralize the charge on growing particles, aggregation of particles occurs with the formation of chains, and ultimately, a three-dimensional gel network is developed. Thus the gel or sol formation depends on the medium containing silicic acid monomer. Black liquor contains silica in the form of sodium silicate and due to high salt concentration gel formation is favored.

4.2 *Effect of Black Liquor Desilication on Chemical, Physical, and Combustion Properties*

Black liquor is one of the highest ash-containing fuels used as an internal fuel in the paper industry and constitutes about 40% of the energy requirements for process operations. The properties of black liquor are important during its processing in chemical recovery plants [39]. Therefore, it is very vital to understand black liquor properties for designing and efficient operation of a chemical recovery boiler. In this section, the effect of the desilication process on commonly used wheat straw and bamboo black liquors is presented before and after carbonation with flue gas.

4.2.1 Desilication of Wheat Straw and Rice Straw Black Liquors

The results of desilication in black liquor collected from wheat straw and rice straw-based mills are presented in this section for studying changes in properties by analyzing various parameters. The physicochemical characteristics of wheat straw black liquor used for the studies are shown in Tables 8 and 9.

It is clear from the results that wheat straw black liquor contains a high amount of silica, which causes problems in the smooth functioning of the chemical recovery system. This high amount of silica in wheat straw black liquor makes recovery of chemicals very difficult and does not allow lime reburning. The viscosity of the wheat straw black liquor is also on the higher side and black liquor becomes unstable above 65% total solids concentrations. The high viscosity of black liquor also causes numerous problems in the evaporation section.

Table 8 Physico-chemical characteristics of wheat straw black liquor

Parameter	Value
pH at 30 °C	12.5
RAA, g/l as NaOH	4.5
Silica as SiO_2, g/l	3.5
Inorganics as NaOH, % *w/w*	36.7
Organics, % *w/w*	63.3
Swelling volume ratio (SVR), ml/g	9.0
Calorific value, cal/g	2797
Carbon as C, % *w/w*	28.5
Hydrogen, % *w/w*	3.10
Nitrogen as N, % *w/w*	0.23
Temperature of ignition, °C	590

Table 9 Viscosity of wheat straw black liquor

Total solids, % *w/w*	Viscosity at 90 °C, mPa.s
50	79
55	199
60	446.7
65	Could not be determined

The impact of carbonation on black liquor viscosity at different pH levels was also studied. This is important to optimize the conditions for desilication of wheat straw black liquor and the results of desilication of black liquor are shown in Table 10. The results clearly reveal that the optimum pH for desilication is 9.9–10.0. The residual silica level is 0.25 g/l which is close to wood black liquor. The desilication was more than 90% with easily filtrable desilicated black liquor without lignin precipitation and moderate foam generation.

The desilication trials conducted on rice straw black liquor are presented in Table 11.

Table 10 Results of desilication trials

Black liquor characteristics			
Before Desilication		After Desilication	
pH	SiO_2, g/l	pH	SiO_2, g/l
12.5	3.5	10.4	3.0
		10.2	0.7
		10.1	0.6
		10.0	0.35
		9.8	0.25

Table 11 Results of desilication of rice straw black liquor at different pH

Black liquor characteristics				Desilication (%)
Before Desilication		After Desilication		
pH	SiO_2, g/l	pH	SiO_2, g/l	
12.0	13.3	10.5	6.9	48.1
		10.4	5.8	56.3
		10.3	1.5	88.7
		10.2	1.0	92.5
		10.1	0.7	94.7
		10.0	0.4	97.0
		9.9	0.30	97.7
		9.7	0.30	97.7

Table 12 Realkalisation of rice straw carbonated black liquor

Caustic addition, g/l as NaOH	RAA, g/l as NaOH	
	Calculated	Actual
0.0	0.8	0.8
2.0	2.8	1.3
4.0	4.8	2.7
6.0	6.8	4.2
8.0	8.8	6.0

Realkalization of Desilicated Black Liquors

The degree of desilication is a function of pH. During carbonation, the pH of the black liquor drops down to 10.0. The residual active alkali of the black liquor drops during carbonation. Carbonated black liquor has an RAA level below 1.0 in most cases. This leaves the lignin in an unstable state. Concentration by evaporation of such liquor causes early viscosity rise and precipitation, long before the concentration required for self-sustained combustion is achieved [40–42]. For stability of black liquor certain level of residual active alkali is essential in black liquor. Realkalsation of carbonated black liquor by the addition of different dosages of alkali is carried out in order to make the black liquor stable. Around 4.0 gpl of caustic is required to check the precipitation during evaporation.

Table 12 shows the effect of caustic addition on RAA of carbonated rice straw black liquor. The study shows the difference between the anticipated (calculated) RAA after addition of caustic and the actual RAA levels achieved after carbonation.

Impact of Desilication on Wheat and Rice Straw Black Liquor Properties

The chemical recovery and thermal efficiency of silica-rich non-wood black liquor chemical recovery plants is low as compared to wood-based mills. The primary reason is the nature of raw material and its effect on black liquor properties like organic to inorganic ratio etc. These components ultimately influence the processing of black liquor in the chemical recovery section. The nature of black liquor changes significantly after desilication. Therefore physico-chemical, rheological, and combustion properties of black liquor were examined after desilication and compared with original black liquor to understand the impact of desilication on the characteristics of the non-wood black liquors.

Physico-chemical Characteristics of Desilicated Black Liquor

Desilicated black liquor was analyzed for physico-chemical properties and the same were compared with original black liquor. The results are shown in Table 13. The results shown in Table 13 have revealed that there is no significant change in the physicochemical properties of wheat straw black liquor after desilication. Silica has gone down to 0.25% as against 3.2 in original black liquor.

Improvements in Thermal Properties of Desilicated Black Liquor

The thermal properties of black liquor are important for the efficiency of the chemical recovery boiler. Swelling volume ratio, calorific value, and Temperature of Ignition were determined before and after desilication. The results are shown in Table 14. The results shown in Table 14 revealed that there is a marginal increase in calorific value. The temperature of ignition has gone down to 575 from 590 °C. The swelling Volume Ratio (SVR) increased from 9 to 15 ml/g after desilications. Figure 10

Table 13 Physico-chemical characteristics of wheat straw black liquor before and after desilication

Parameter	Value	
	Before Desilication	After Desilication
pH at 30 °C	12.5	12.3
RAA, g/l as NaOH	4.5	3.5
Silica as SiO_2, g/l	3.2	0.25
Inorganics as NaOH, % *w/w*	36.7	36.8
Organics, % *w/w*	63.3	63.2
Carbon as C, % *w/w*	28.5	28.1
Hydrogen, % *w/w*	3.10	2.99
Nitrogen as N, % *w/w*	0.23	0.28

Table 14 Improvement in thermal properties of desilicated wheat straw black liquor

Parameter	Value	
	Before Desilication	After Desilication
Swelling volume ratio (SVR), ml/g	9	15
Calorific value, cal/g	2797	2813
Temperature of ignition (Tig), °C	590	575

Fig. 10 Improvement in swelling volume ratio after desilication

shows the improvement in swelling characteristics of wheat straw black liquor after desilication. An increase in the Swelling Volume Ratio and lower Temperature of ignition value is a pointer towards better-burning behavior of desilicated black liquor. A marginal increase in the calorific value will improve thermal energy recovery from black liquor.

Improvement in Viscosity of Desilicated Black Liquor

Black liquor is a colloidal system, and its viscosity depends largely upon the colloidal stability of lignin, and macromolecules present. Higher pH and the presence of free alkali ensure that lignin remains in colloidal form. On carbonation, free alkali is

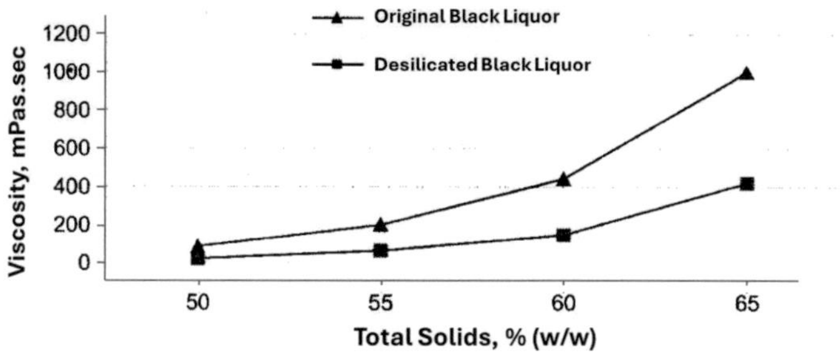

Fig. 11 Viscosity of original and desilicated wheat straw black liquor

Table 15 Viscosity of rice straw black liquor before and after desilication

Total solids, %*w/w*	Viscosity, mPa.s at 90 °C	
	Original black liquor	Desilicated/realkalised black liquor
35	33	8
40	56	12
45	112	24
50	316	75
52	730	130
55	Could not be concentrated as the liquor becomes unstable and got precipitated	194

almost destroyed, and lignin becomes colloidally unstable. Therefore, the addition of caustic is necessary to retain the fluidity of the liquor at higher concentrations. The viscosity of the re-alkalized desilicated black liquor and original black liquor was determined at increasing dry solids concentrations. Figure 11 represents the viscosity curve of original and desilicated wheat straw black liquor showing the influence of silica on viscosity. The figure shows that a significant reduction in viscosity is achieved after desilication of wheat straw black liquor. High viscosity values of black liquor are generally the function of hemicellulose content in the black liquor. As desilication is carried out at a temperature much below the hemicellulose degradation temperature, it is obvious that the removal of silica caused a marked reduction in black liquor viscosity.

The results of pilot plant trials on viscosity reduction of desilicated and nondesilicated rice straw black liquors are presented in Table 15.

4.2.2 Desilication of Bamboo Black Liquor

Bamboo black liquor was carbonated at different pH levels to optimize the pH for maximum desilication efficiency. Physico-chemical analyses of bamboo black liquors before and after carbonation are presented in Table 16.

The results of the carbonation of bamboo black liquor are depicted in Table 17. Thus, it is evident that silica precipitation starts at a pH of 10.5 and 90% desilication is achieved at a pH of 10.1. Results show that residual silica content is almost same after pH level of 10.0. This indicates that carbonation of black liquor beyond pH 10.1 does not facilitate silica precipitation.

The above results indicate that optimum pH range for precipitation of silica in bamboo black liquor is 10.1–10.0. Residual silica content in black liquor is 0.4 g/l in this pH range. It is difficult to remove this residual silica by carbonation.

Table 16 Physico-chemical characteristics of bamboo black liquor before and after carbonation

Parameter	Original black liquor	Desilicated/realkalised black liquor
pH at 25 °C	12.8	12.3
Total solids, % *w/w*	19.5	19.2
R.A.A., g/l as Na_2O	7.0	4.0
Silica as SiO_2, % *w/w*	2.15	0.20
Silica as SiO_2, g/l	4.2	0.4
Inorganics as NaOH, % *w/w*	36.8	38.2
Organics, % *w/w*	63.2	61.8
Mixed oxide as R_2O_3	0.1	0.1
Carbon as C, % *w/w*	31.7	31.3
Hydrogen as H, % *w/w*	3.40	3.4
Nitrogen as N, % *w/w*	0.14	0.05

Table 17 Results of desilication of bamboo black liquor at different pH

Black liquor characteristics				Desilication, %
Before Desilication		After Desilication		
pH	SiO_2, g/l	pH	SiO_2, g/l	
12.8	4.2	10.5	2.1	48.8
		10.4	1.5	64.3
		10.3	1.2	71.4
		10.2	0.8	80.9
		10.1	0.5	88.9
		10.0	0.4	90.5
		9.8	0.4	90.5

Table 18 shows the effect of carbonation in RAA of the bamboo black liquor. As the pH of the black liquor is lowered due to controlled carbonation, RAA depletion takes place. Hence, it is clear that, with the onset of desilication, RAA depletes very fast as desilication proceeds.

Caustic alkali is added prior to the evaporation of the desilicated black liquor to maintain the colloidal nature of the black liquor. A study was conducted to find out the depletion of RAA after carbonation and then to reinstate the RAA level by the addition of caustic in the carbonated black liquor, prior to evaporation in MEE. The results of this study are presented in Table 19. Table 19 shows the effect of caustic addition on RAA of carbonated bamboo black liquor. The difference between the anticipated (calculated) RAA after the addition of caustic and the actual RAA levels achieved after carbonation is presented in Table below. The results of bamboo black liquor viscosity reduction after addition of different dosage of NaOH are shown in Table 20.

Impact of Desilication on Bamboo Black Liquor Properties

Black liquor properties of bamboo black liquors were evaluated before and after desilication. The Thermogravimetric curves of the desilicated bamboo black liquor

Table 18 Effect of carbonation on RAA of bamboo black liquor

Initial pH	pH after carbonation	RAA, g/l as Na_2O	% Desilication
12.8		7.0	–
	10.9	3.4	–
	10.6	2.3	–
	10.5	2.0	48.8
	10.4	1.6	64.3
	10.3	1.0	71.4
	10.2	0.8	88.1
	10.1	0.6	88.9
	10.0	0.5	90.5

Table 19 Realkalisation of bamboo carbonated black liquor

Caustic addition, g/l as Na_2O	RAA, g/l as Na_2O	
	Calculated	Actual
0.0	1.0	1.0
2.0	3.0	1.2
4.0	5.0	2.8
6.0	7.0	4.1
8.0	9.0	5.7

Table 20 Viscosity of bamboo black liquor after addition of different dosages of alkali (in m.Pa.S)

Total solids, % *w/w*	Original black liquor viscosity (RAA 7.0 g/l)	Desilicated black liquor viscosity, mPa.s			
		NaOH added, g/l as Na_2O			
		2.0	4.0	6.0	8.0
60	55	Black liquor is unstable beyond 50% Total solids	88	45	48
65	128		204	96	120

Table 21 Thermal properties of bamboo black liquors

Parameters	Original black liquor	Desilicated black liquor
Swelling volume ratio, ml/g	29	40
Gross calorific value, cal/g	2940	2890
Drying range, °C	80–180	80–180
Pyrolysis range, °C	240–415	266–425
Weight loss during pyrolysis, %	28.6	29.3
Combustion range, °C	590–645	579–635
Weight loss during combustion, %	20.7	20.6
Temperature of ignition, °C	639.4	632

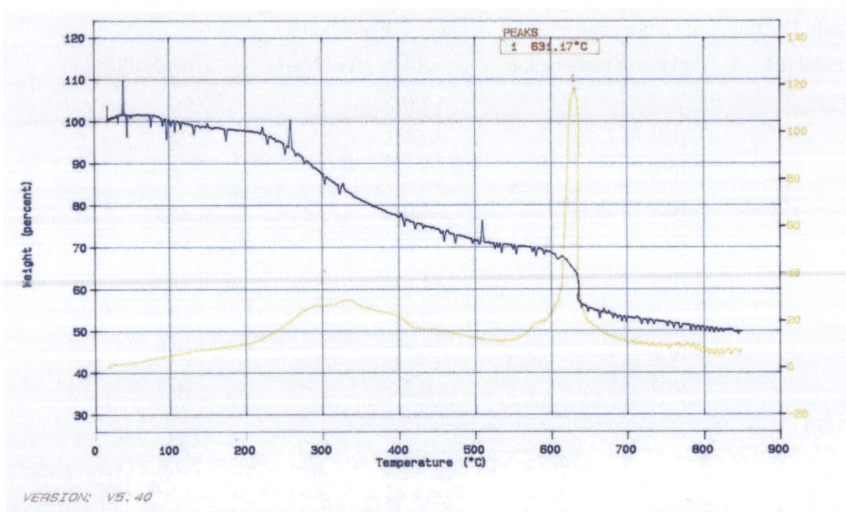

Fig. 12 TGA curve of desilicated bamboo black liquor

are shown in Fig. 12. Based on the TGA analysis, the analysis of the thermal behavior of desilicated and non-desilicated bamboo black liquors is presented in Table 21.

The following observations are made on the desilication of bamboo black liquor.

- The original bamboo black liquor was found to be highly colloidally stable and no lignin co-precipitation was observed during its controlled carbonation.
- Foam generation during gas–liquid interaction was found to be moderate and lie within highly manageable limits.
- Highly encouraging results were obtained during silica settling studies on carbonated black liquor. After three hours of settling the sludge volume was close to 20% of total liquor volume and supernatant black liquor had less than 20 ppm suspended solids.
- The viscosity of the desilicated and realkalized black liquor was fairly close to the original black liquor.
- There was only a marginal reduction in the calorific value of realkalized bamboo black liquor, due to an increase in the concentration of inorganics during realkalisation.
- There was a remarkable improvement in the swelling volume ratio (SVR) of carbonated and realkalized black liquor when compared to the original black liquor, which is a pointer toward better combustibility of desilicated black liquor.

4.3 Analysis of Silica Sludge Separated from Black Liquor

Silica sludge obtained in different desilication trials was collected after washing with hot water. Precipitated silica was analyzed for various parameters, It has a whitish appearance, having a brightness of 68.4% ISO. It had about 82% ash, and on an ash basis, it had 95% silica. On an O.D. sludge basis, silica content is 79.8%. Only traces of lignin are found in the silica sludge. The results are depicted in Table 22. Wheat straw silica may find application in paints, rubber, pesticides, and the tire industry. Figure 13 shows precipitated silica from wheat straw black liquor before and after hot water washing.

Table 22 Analysis of wheat straw silica

Parameter	Value
Ash, %w/w	82.2
Silica, %w/w, ash basis	95.0
Silica, %w/w, sludge basis	79.8
Lignin, %w/w	0.1
Brightness, %ISO	68.4

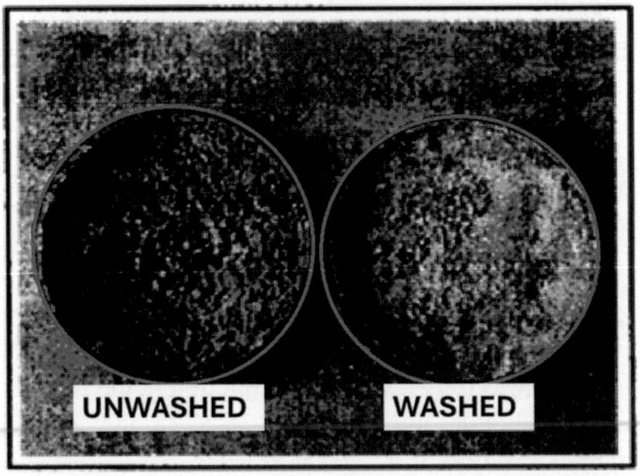

Fig. 13 Silica separated from wheat straw black liquor

5 Conclusions

The thermal treatment of black liquor is a promising technology, especially for bagasse-based mills where higher black liquor viscosities have been one of the major problems restricting the firing of black liquor at high dry solids, thereby resulting in poor thermal efficiency of the recovery boiler. Studies have shown that it is not possible to concentrate bagasse black liquors above 60% solids concentration in conventional chemical recovery systems. However, the retrofitting of the chemical recovery system with a thermal treatment unit can help to achieve higher soilds concentration by reducing bagasse black liquor viscosity.

The thermal treatment trials conducted on a pilot scale with bagasse and eucalyptus black liquors have shown a drastic reduction in viscosity and it can enable the bagasse black liquor to evaporate up to or above 70% solids concentration after thermal treatment. After upscaling the pilot scale process to the commercial scale it was observed that retrofitting of the thermal treatment system in the chemical recovery island could improve the energy efficiency by more than 10% by generation of additional steam to the tune of 0.35 t/t solids. Besides the benefits of improvement in the energy efficiency of the chemical recovery system, the process also results in an improvement in the environmental situation of the mill by way of reduced $S0_2$ emission by removing the direct cascade evaporator from the recovery system.

The thermal treatment system is a safe and easy process and can be easily retrofitted in existing chemical recovery systems. It should be integrated as a part of recovery installation in existing and upcoming bagasse-based mills. The desilication of black liquor is also an important technology for bamboo and straw-based mills to run the chemical recovery operations in a smooth way as the presence of a high amount of silica in bamboo and wheat straw black liquor causes numerous problems in chemical recovery installations. The desilication of black liquor has shown that silica can be effectively removed from wheat straw and bamboo black liquors to the tune of 90%. Detailed black liquor analysis after desilication has shown that the burning and rheological properties of black liquors are improved. This can help in improving the recovery of chemicals and enhance the thermal efficiency of the recovery boiler. Residual silica content in the range of 0.2–0.3 gpl, after desilication, is very close to wood black liquor and with this silica content in non-wood black liquors, efficient lime kilns can be installed in bamboo and straw-based mills.

References

1. Casey JP (1980) Textbook on "pulp and paper chemistry and chemical technology", vol 1, 3rd edn, p 456
2. Hubbe M, Alén R, Paleologou M, Kannangara M, Kihlman J (2019) Lignin recovery from spent alkaline pulping liquors using acidification, membrane separation, and related processing steps: a review. BioRes 14(1):2300–2351
3. Judt M (1991). Desilication problem can be overcome. Pulp Pap Int 33(6); Judt M (1994) UNIDO offers a helping hand to developing countries. Pulp Pap Int 36(8):42–43
4. Bajpai P (2018) Pulping fundamentals. In: Biermann's handbook of pulp and paper. Elsevier, pp 295–351. https://doi.org/10.1016/B978-0-12-814240-0.00012-4
5. Kulkarni AG, Pant R, Panda A (1993) Treatment of bagasse black liquor. IPPTA 5:4–10
6. Huang G, Shi J, Langrish T (2007) NH_4OH-KOH pulping mechanisms and kinetics of rice straw. Bioresour Technol 98:1218–1223. https://doi.org/10.1016/j.biortech.2006.05.002
7. Yuan Z, Kapu NS, Beatson R, Chang XF, Mark Martinez D (2016) Effect of alkaline pre-extraction of hemicelluloses and silica on kraft pulping of bamboo (*Neosinocalamus affinis Keng*). Ind Crops Prod 91:66–75. https://doi.org/10.1016/j.indcrop.2016.06.019
8. Wilder HD, Daleski EJ (1964) Kraft pulping kinetics. I. Literature review and research program. Tappi 47:270–275
9. Fergus BJ, Goring DAI (1970) The location of guaiacyl and syringyl lignins in birch xylem tissue. Holzforschung 24:118
10. Kawamura I, Higuchi T (1964) Comparative studies of milled wood lignins from different taxonomical origins by infrared spectroscopy. Chimie et Biochimie de la Lignine de la Cellulose et des Hemicelluloses 439–456
11. Sarkanen KV, Ludwing CH (1971) Lignin occurance, formation, structure, and reactions. Wiley – Interscience & Son, Inc., N.Y., pp 1–65
12. Jin Z, Katsumata KS, Lam TBT, Iiyama K (2006) Covalent linkages between cellulose and lignin in cell walls of coniferous and nonconiferous woods. Biopolymers 83(2):103–110. https://doi.org/10.1002/bip.20533
13. Wood JR, Goring DAI (1971) Distribution of lignin in stem wood and branch wood of Douglas fir. Pulp Pap Mag Can. 72:61–68
14. Higuchi T (1990) Lignin biochemistry: biosynthesis and biodegradation. Wood Sci Technol 24:23–63
15. Bisht M (1996) Characterization of cell wall component of wood & bark of *eucalyptus toreticornis* and their behavior during alkaline delignification. D.Phil thesis, Garhwal University
16. Creighton RHJ, Gibbs RD, Hibbert H (1944) Studies on lignin and related compounds. LXXV. Alkaline nitrobenzene oxidation of plant materials and application to taxonomic classification. J Chem Soc 66:32–37
17. Higuchi T, Ito Y, Kawamura I (1967) P-hydroxyphenylpropane component of grass lignin and role of tyrosine-ammonia lyase in its formation. Phytochemistry 6(6):875–881. https://doi.org/10.1016/S0031-9422(00)86035-5
18. Smith D (1955) Ester groups in lignin. Nature 176:267–268. https://doi.org/10.1038/176267a0
19. Gierer J (1970) The reactions of lignin during pulping. A description and comparison of conventional pulping processes. Svensk papperstidn 73(18):571–596

20. Gierer J, Lindeberg OTTO (1980) Reactions of lignin during sulfate pulping. Part XIX. Isolation and identification of new dimers from a spent sulfate liquor. Acta Chem Scand B 34:161–170
21. Marton J (1971) Reactions in alkaline pulping. In: Sarkanen KV, Ludwig CH (eds) Lignins: occurrence, formation, structure, and reactions. Wiley Interscience, New York, pp 639–694
22. Procter AR, Yean AQ, Goring DAI (1967) Topochemistry of delignification in pulping: secondary wall delignified faster than middle lamella. Pulp Pap Mag Can 68(9):T445
23. Adler E (1977) Lignin chemistry—past, present and future. Wood Sci Technol 11(3):169–218
24. Gierer J (1980) Chemical aspects of kraft pulping. Wood Sci Technol 14:241–266. https://doi.org/10.1007/BF00383453
25. Mische GE (1972) Über das Verhalten des lignins bei der Alkalikochung. VIII. Isomerisierung der Phenolatanionen von erythro- und threo-Isoeugenolglykol-beta-(2-methoxyphenyl)-äther über ein Chinomethid. Acta Chem Scand 26:4137–4142
26. Gierer J, Smedman LA (1965) The reactions of lignin during sulphate cooking. Part VIII. The mechanism of splitting of beta-arylether bonds in phenolic units by white liquor. Acta Chem Scand 19:1103–1112
27. Wilder HD, Daleski EJ (1965) Kraft pulping kinetics. II. Delignification rate studies. Tappi J 48:293–297
28. Gierer J, Lindberg O (1980) Reactions of lignin during sulfate pulping. Part XIX. Isolation and identification of new dimers from a spent sulfate liquor. Acta Chem Scand B 34:161–170
29. Shemyakin MM, Shchukina LA (1956) Oxidative-hydrolytic splitting of carbon–carbon bonds of organic molecules. Q Rev Chem Soc 10(3):261–282. https://doi.org/10.1039/qr9561000261
30. Sarkanen KV, Hurtiford BF, Johnson LN, Gardner HS (1970) Kraft odor. Tappi 53:766–783
31. Kulkarni AG, Mathur RM, Panda A (1989) Nature of spent liquors from pulping of non-wood fibrous raw material. In: Tappi proceedings of the conference of international symposium on wood and pulping chemistry, Raleigh (USA), pp 485–492
32. Dlxlt AK, Mathur RM, Kulkarni AG (2006) Desilication of wheat straw black liquor, IPPTA convention issue, pp 75–79
33. Jain RK, Dixit AK, Nair RK, Mathur RM, Kulkarni AG (2000) Improved energy efficiency through thermal treatment of black liquor-a pilot scale experience. IPPTA 12:47–52
34. Kulkarni AG, Pant R, Panda A (1991) Breakthrough in desilication technology—new opportunities in processing of non-wood spent pulping liquors. In: Proceedings 6th international symposium on wood and pulping chemistry 1991. Melbourne, Australia
35. Dixit AK, Thapliyal BP, Jain RK, Mathur RM (2010) Desilication of bamboo & straw black liquor. IPPTA 22(3):121
36. Isono Z, Ono K (1968) Studies on the desilicification from high silica-containing kraft liquors part III. Desilicification from Kraft black liquor by aluminium sulfate. Nippon Någeikagaku Kaishi 42(1). https://doi.org/10.1271/nogeikagaku1924.42.24
37. Isono Z, Ono K (1967) Studies on the desilicification from high silica-containing kraft liquors part I. Desilicification from black liquor by carbon dioxide treatment. Nippon Någeikagaku Kaishi 41(5). https://doi.org/10.1271/nogeikagaku1924.41.5_220
38. Aujla MI, Ishtiaq-Ur-Rehman I-U-R, Javaid A (2007) Mechanism of silica precipitation by lowering pH in chemi-thermomechanical pulping black liquors. In: Proceedings of the 1st WSEAS international conference on computational chemistry
39. Wallmo H, Richards T, Theliander H (2007) Lignin precipitation from kraft black liquors: kinetics and carbon dioxide absorption. Paperi Ja Puu/Pap Timber 89(7–8)
40. Kulkarni AG, Kolambe SL, Mathur RM, Pant R (1984) Studies on desilication of bamboo kraft black liquor. IPPTA 21(1):37–44
41. Söderhjelm L (1986) Viscosity of strong black liquor. Paperi ja Puu-Paper and Timber, 68(9):642–652
42. Soderhjelm L, Sagfor IPE (1992) Relationship between the viscosity and composition of black liquor. Int Chem Recov Conf Book 2:513–519

Development of Modern Materials from Lignosulphonates

Saakshy and Rahul Mishra

Abstract The end use of lignin is dependent on the structure which in turn is affected by the raw material and the pulping method or the extraction process. Lignosulphonates are organic polymers extracted from the black liquor of sulphite pulping process. In this chapter, the detailed utilization of ligno-sulphonates has been mentioned including dust control, lubricants, antioxidants, fire retardants, ceramics, chelating agents, binders, emulsifiers, fillers, dispersants, sodium ion batteries, cathode for lithium ion batteries, lead acid battery, carbon fibre, fertilizers, adhesives, enhancer of thermal behavior, animal species and construction material. The development of modern materials from lignosulphonates reduces the dependency on petroleum based materials and open avenues for greener options utilizing renewable resources from lignosulphonates.

Keywords Lignosulphonates · Black liquor · Sulphite pulping · Lignin

1 Introduction

Lignin, a natural biopolymer on the earth, [1] has three types when produced from paper industry, namely soda lignin, kraft lignin and sulphite lignin [2]. Lignin is a phenyl propanoic monomer [3] and largest reservoir of aromatic compound. Lignin result from the oxidative coupling of, primarily, 4-hydroxyphenylpropanoids [4]. There are three types of lignin according to the plant species: softwood, hardwood and grass lignin. Softwood lignin consists exclusively of coniferyl alcohol, hardwood lignin consists mainly of coniferyl alcohol and sinapyl alcohol, grass lignin has three types of monomers (coniferyl, sinapyl and p-coumaryl alcohol) [5]. The novel pulping methods includes steam explosion and organosolv pulping. The studies conducted by some researchers showed that lignin has anti-tumour activity due to the presence of low molecular weight monomers [6]. Lignin utilization is

Saakshy (✉) · R. Mishra
Kumarappa National Handmade Paper Institute, Jaipur, Rajasthan, India
e-mail: saakshyagarwal@gmail.com

S. Sundaramurthy et al. (eds.), *Black Liquor Waste Management in the Pulp and Paper Industry*, Sustainable Materials and Technology,
https://doi.org/10.1007/978-981-96-4237-3_5

still low if compared with the annual production for making value added products. Lignosulphonates are by-products of the pulp and paper industry form the pulping of lignocellulosic raw material adopting sulphite pulping. Currently, lignosulphonates contribute to 90% of the total market of commercial lignin in the form of calcium lignosulfonate, sodium lignosulfonate and magnesium lignosulfonate [7].

2 Preparation of Lignosulphonates

The sulfite process is a chemical pulping method in which the raw material is treated with a solution of sulfite and bisulfite ions which cleave the bond between the cellulose and lignin components of the lignocellulose. The raw material is cooked with sulphite as cooking liquid in the sulphite pulping process [8]. In alkaline sulphite pulping process, sodium hydroxide and sodium sulphite [10] are used for delignification while in neutral sulphite pulping process, sulphite liquor is prepared by the dissolution of sulphur dioxide in sodium carbonate. The residual lignin in sulphite process is sulphonated. Lignosulphonates are generated as a by-product during sulphite pulping (Fig. 1). The efficient and eco-friendly methods of extracting lignin from lignocellulosic biomass has been given in the previous studies [9].

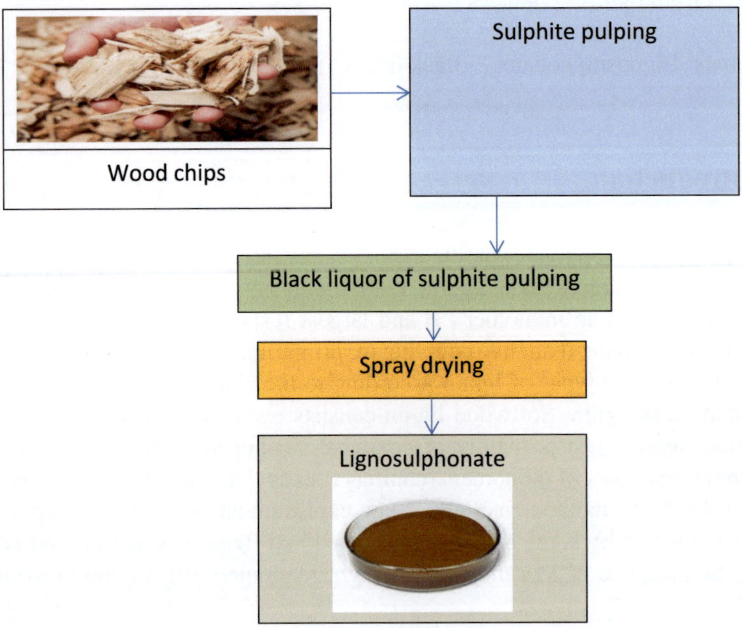

Fig. 1 Lignosulphonate extraction process

3 Utility of Lignosulphonates

It is widely known that lignin is a renewable source of phenolic compounds and there is growing interest in using it as a substitute for petroleum based products; however, still there is low utilization of lignin for making value added products. There is great potential of lignin to be converted to high value-added phenolic platform chemicals. Lignosulphonates derived from the sulphite pulping process, account for 90% of the total market of commercial lignin [11]. Lignin is mostly used as an energy substance, lignin-based carbon fibers, due to the presence of phenolic hydroxyl groups and aliphatic hydroxyl groups in the lignin structures [12]. Lignosulphonates already exists in market for making binding agents for fuel, pellets as feed additive, dispersant, as an analogue of humic acid, feeds, ensilage and flame retardants due to its anti-oxidant effect, sunscreens and biopesticides due to their UV absorbance (Fig. 2).

3.1 Dust Control

The sodium and calcium lignosulphonate was found to reduce the dust erosion of red sand [13]. Lignosulfonate is applied for dust control by binding particles of road surface together. After application, water evaporates from the lignin on drying, resulting in trapping of dust particles in high-viscosity sticky mass of lignin. Over time, the lignin becomes completely "insoluble due to solar heating", creating a tough, water-resistant, and durable dust-controlled surface. Lignosulfonate can be

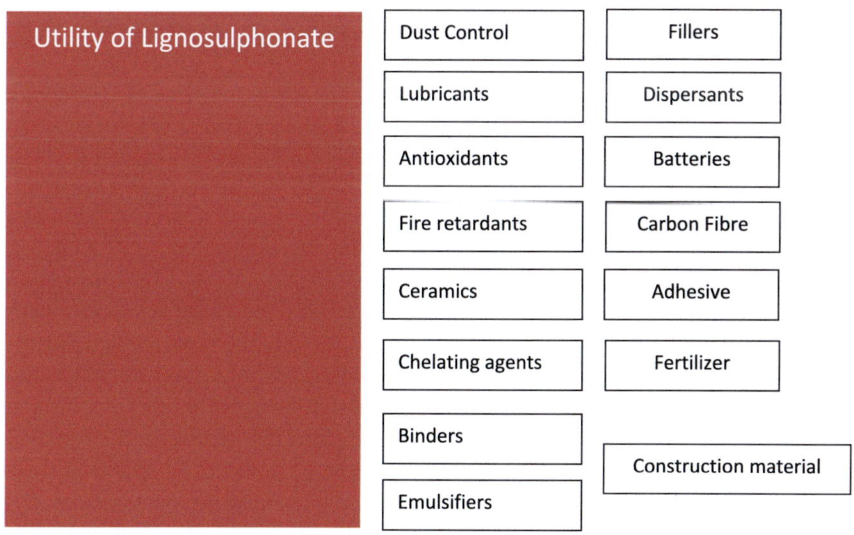

Fig. 2 Various utility of lignosulphonates

used on various surfaces such as feed lots, orchards, county, roads, construction areas, tree and flower nurseries. Lignosulfonate is considered "environmentally safe, non-corrosive, and non-toxic to nearby plants and wildlife" [14].

3.2 Lubricants

Lignosulfonates have been used as lubricants in various industrial applications. For example, they have been used to lubricate ingot molds in steel production, where they generate only a small amount of light smoke when teemed into the molds [15]. Lignosulfonates have also been used in the production of lubricating greases, where they act as a thickening agent and improve the greases' performance [16]. In the oil industry, lignosulfonates are used as deflocculants to prevent the coagulation of solids [17]. They have been successfully used in lime-based and brine-based drilling fluids or drilling muds [18].

3.3 Antioxidants

Lignosulfonates have been studied for their antioxidant properties due to the presence of phenolate and carboxylate groups [19, 20]. The previous studies have shown that lignosulfonates have antiviral properties which may stem from its polyanionic nature [21]. Studies also been conducted on lignosulfonates for their potential "as an antioxidant in the production of lubricating greases" [22]. However, more research is needed to fully understand the antioxidant capabilities of lignosulfonates.

3.4 Fire Retardants

Lignosulfonates have been studied for their potential as fire retardants. Research has shown that lignosulfonates can improve the thermal behavior of materials such as rattan, lead acid batteries, other energy storage applications. Their thermal stability and fire resistance make them a valuable additive in improving the performance of different materials in various applications including areca particle boards, making them more fire-resistant and thermally stable [23, 24]. Lignosulfonates have been found to act as char-forming agents after the decomposition process forming a coherent char layer that slows down heat transfer [25]. Additionally, lignosulfonates have been shown to enhance flame retardancy property as indicated by an increase in the Limiting Oxygen Index with increase in lignosulfonate content [26]. These findings suggest that lignosulfonates hold promise as fire retardants in various applications.

3.5 Ceramics

Lignosulfonates are used in ceramics to increase the dry strength, reduce the amount of plastic clay, improve mobility, and increase yield in ceramic items resulting in also a reduction in the costs [27]. Studies have been conducted on the addition of lignosulphonates in ceramic items at varying concentrations typically ranging from 0.1 to 2.0%. In pressed tiles, it has been found that 1% lignosulfonate has double the mechanical strength. Lignosulfonates are also used to reduce shrinkage while maintaining after-forming mechanical strength. Lignosulphonates are water soluble organic surfactants [27]. The calcium lignin gets adsorbed on the surface of raw material particles increasing the liquidity of mud [28].

3.6 Chelating Agents

Lignosulphonates contain functional groups that display a chelating property, capturing polyvalent metal ions and forming complex compounds [29]. Sodium lignosulphonate is found to be an "effective coagulant owing to its ability to readily complex with positively charged colloidal particles at acidic pH" [30]. Lignosulphonates are versatile chelating agents with applications in agriculture, water treatment, soil improvement, and various industrial processes. The chelating properties of lignosulphonates make them valuable in ensuring the availability and stability of micronutrients in different environments. Ligno-sulphonates have different ionic charges as chelating agent for trace elements, which supply micronutrients in agriculture. Lignosulphonates are also used as soil application agents [31]. In addition to their use in agriculture, lignosulphonates are used as chelating agents in various other applications.

3.7 Binders

Lignosulphonates works as binders in various end applications including particle boards, wall paper, linoleum flooring, coal briquettes, and roads [32]. Lignosulphonates help in binding coal powder and refinery coke into regular sizes as well as reduce the cost of production by decreasing the amount of feed powder and lower the anti-back ratio [33]. There are several patents which shows role of lignosulphonates as a binder for pyrolysis of carbon black. The binding properties of lignosulphonates contribute to the strength of carbon black pellets [34]. Overall, lignosulphonates are used as binders in various industries, including construction materials, ceramics, and wood-based materials, due to their adhesive properties and versatility in different applications.

3.8 Emulsifiers

Lignosulphonates stabilize oil-water emulsion mechanism by a combination of electrostatic and steric stabilization [35]. Lignosulphonates are adsorbed at the oil-water interface establishing a high order electro kinetic charge which helps in stabilization of the emulsion [36]. Lignosulphonates have been found to yield emulsions that are stable over months and longer, finding application in agrochemical formulations [35]. Lignosulphonates are biobased surfactants due to its amphilic nature which makes it applications in various applications including stabilizers, plasticizers, dispersants and agrochemical formulations [37].

3.9 Fillers

Lignosulphonates are used as fillers in ceramic tiles, resins to fiber boards, ceramics, rubber, casting sand, fodder pellets and wood based materials. Lignosulphonates are used as fillers and improve strength properties of rubber compounds [38].

3.10 Dispersants

Lignosulphonates are versatile dispersants that stabilize dispersions, emulsions and are found in applications such as concrete formulations, ceramics, coatings, drilling fluids, and more [39–41]. In the construction industry, lignosulphonates are often used as dispersants in concrete to improve its workability and reduce the amount of water needed for mixing, resulting in stronger and more durable concrete [42]. Additionally, lignosulphonates are used as dispersants in the production of ceramics resulting in energy savings [43]. Overall, lignosulphonates are valued for their ability to disperse or distribute solid particles in various liquid media making them useful in various industrial applications.

3.11 Batteries

Sodium lignosulfonate has been demonstrated as a promising binder for hard carbon anodes in sodium-ion batteries. It has been shown to exhibit high performance, stability, and cost-effectiveness in this application. Research has indicated that in sodium-ion batteries, lignosulfonate shows better performance with high capacity retention and stability [44]. Additionally, sodium lignosulfonate has been tested as a binder for hard carbon anodes in sodium-ion batteries and compared with lignin, showing its potential as a green next-generation battery electrode binder [45]. Furthermore, sodium lignosulfonate has been utilized as an electrolyte additive to stabilize zinc deposition and improve the lifespan of aqueous zinc-ion batteries [46], the potential of lignosulfonates in enhancing the performance and sustainability of sodium-ion and zinc-ion batteries [47]

Lignosulfonate has been explored as a potential material for use in lithium-sulfur batteries as high performance cathode [48]. The electrochemical stability of cathodes derived from lignosulfonate has been found to be superior compared to their lithium-ion counterparts [49]. Additionally, lignosulfonate has been employed as a binder for hard carbon anodes in sodium-ion batteries and showed high capacity retention and stability [44].

Lignosulphonate has been studied as a binder for lithium-ion batteries. The studies conducted on utilization of lignosulphonate as a binder for hard carbon anodes in sodium-ion batteries, demonstrated high performance, stability, and cost-effectiveness [44]. Lignosulphonate has also been explored as a potential binder for chitin-derived carbon materials as a superior anode for lithium-ion batteries [50]. Overall, it can be concluded that lignosulphonates are a sustainable and high-performance binder for lithium-ion batteries [51].

Lignosulphonates have been studied for their potential use in lead-acid batteries, specifically for "the negative active mass of lead-acid batteries". The previous studies showed that "lignosulphonates extracted from jute sticks have been used to construct the negative active mass of lead acid battery" [7] due to higher solubility in water. "The normalized capacity of the constructed cells" [52] showed better specific charge capacity at lower discharge rates in comparison with standard commercial product [53].

3.12 Carbon Fibre

Lignosulfonates have been used for the production of carbon fibers, which have various applications in industries such as aerospace, automotive, and sporting goods. Lignosulfonates based activated carbon fibers have been developed as "highly efficient adsorption and filtration materials" [54]. These fibers have been used in a wide range of applications, including "wastewater treatment, air purification, and carbon capture". The lignin as an alternative precursor showed "significant reduction in the cost of carbon fiber reinforced polymers and reduction of carbon dioxide emission during carbon fiber production" [55]. Different methods have been used to produce "lignin-based carbon fibers, including melt spinning, solution spinning, electrospinning, melt-blow, centrifugal spinning and barrel spinning" [56]. These methods allow for the creation of carbon fibers with varying properties and applications, such as aqueous zinc-ion batteries [57]. The development of lignosulfonate-based activated carbon fibers and lignin-based carbon fibers [58] demonstrates the potential of these materials in various industries, including textiles, filtration, and energy storage.

3.13 Fertilizer

The calcium lignosulphonate introduced as an "environmentally friendly inhibitor for the selective separation of arsenopyrite and chalcopyrite", "Cu-As separation by froth flotation at low alkalinity" [59, 60]. The lignosulphonates achieved using sulphite pulping of wood pulp are "water soluble polymeric by-products" [61] and can be used as "soil amendments in agriculture". The lignosulphonates acts as "biostimulants in enhancing the action of fertilizers. In soils, they affect the nitrogen and phosphorous cycles, as well as acting as transporters of micronutrients" [43]. Lignosulphonates are widely used in agriculture as fertilizers and soil conditioners. They are used to improve the soil fertility, "soil structure, and water retention", and to promote the growth of "plant foliage, roots, and fruits" [62]. Lignosulphonates are often used in combination with chemical fertilizers such as urea, DAP, MAP, and MKP. They can help stabilize nutrients in the soil, regulate soil pH, and create a favorable environment for microbial activity. Additionally, they aid in the development of chlorophyll, amino acids and sugars in plants by stimulating seed germination and increasing plants' ability to withstand stress and disease. Lignosulphonates are also used as carriers for micronutrients and slow-release fertilizers and improve efficiency of macro-nutrient fertilizer [63]. The addition of lignosulphonate reduces the lime consumption in expansive soil stabilization [64]. Overall, lignosulphonates play a significant role in enhancing soil quality and promoting plant growth in agricultural settings.

3.14 Adhesives

Lignosulphonates have been studied as phenol-based adhesives as an alternative to non-renewable binders in wood-based materials. Lignin is used as a partial substitute in "phenol–formaldehyde or urea formaldehyde resins in wood adhesive" [65]. Magnesium lignosulphonate has been used for manufacturing particle board [66]. Lignosulphonates, as an adhesives finds its place in various applications, such as starch-based adhesive and wood-based materials. Lignosulphonates after modification with carboxylic acids improve the "wet resistance and bonding properties of starch-based adhesives". A previous study showed that enzymatically polymerized lignosulfonates can improve the properties of starch-based adhesives or in the "development of formaldehyde-free adhesives with lignin" and lignosulphonates [67]. These adhesives have shown promising results in terms of mechanical properties and environmental sustainability. Lignosulphonates have been used in the production of particle boards with "low formaldehyde emissions and good mechanical properties". Magnesium lignosulfonate has been used as a key component in the adhesive [66]. Enzymatic synthesis has been used to improve the wet resistance of starch-based adhesives by incorporating lignosulphonates. This strategy has been shown to be a robust method for enhancing the performance of "starch-based

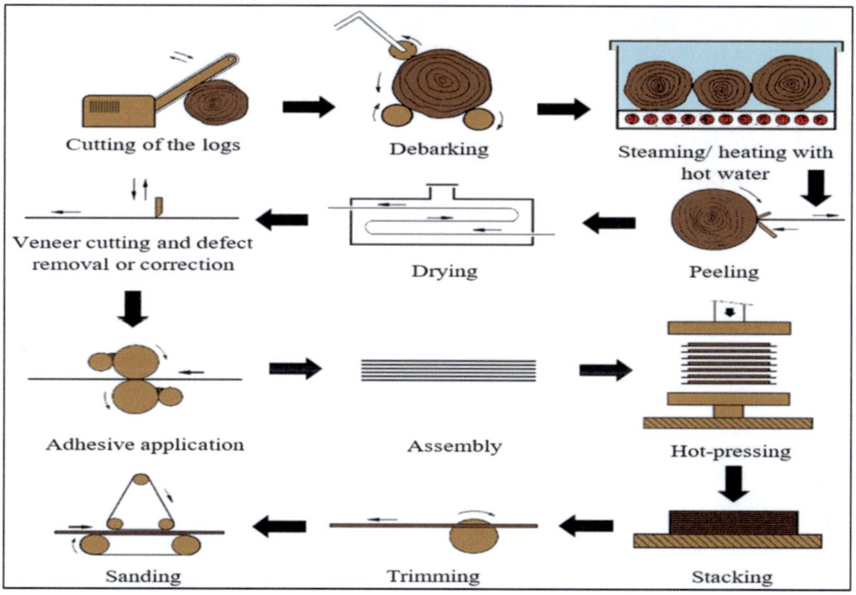

Fig. 3 Utility of lignosulphonate as adhesive in plywood making [70] (*Source*)

adhesives in wet environments" [68, 69]. Overall, lignosulphonates have been used as adhesives in various applications, including starch-based adhesives, wood-based materials, particle boards, and resin products, due to their adhesive properties and versatility in different applications (Fig. 3).

3.15 Construction Material

The lignosulphonates based binders stabilize cohesive soil. [71]. The soil treated with calcium lignosulphonate showed increased "compressive strength, yield stress and the plastic viscosity" [72]. Lignosulphonates are used in concrete, cement slurry, plaster blends, and other construction materials to improve the workability, flowability, and mechanical strength of the final product. Lignosulphonates act as plasticizers or water reducers in concrete, allowing for the production of stronger concrete with less water while maintaining the flow, thus assist in grinding and as a deflocculant in the construction industry (www.transparentmarketresearch.com). They are also used as "fillers and binders in ceramic tiles, resins for fiber boards, casting sand, and fodder pellets. Furthermore, lignosulphonates are employed as dispersants in various construction materials, providing flowability and plasticity to cement and improving impermeability and freeze-thaw performance of concrete". Overall, lignosulphonates play a crucial role in enhancing the performance and quality of construction materials and processes [73] (Table 1).

Table 1 Applications of lignosulphonates in modern materials

S.N.	Application	Products	Form of lignosulphonates/factor for utility
1	Dust control	Construction, feed lots, orchards	Sodium lignosulphonate Calcium lignosulphonate
2	Lubricants	Ingot molds in steel production, lubricating greases	Alkali metal lignosulphonate
3	Antioxidants	Lubricating greases	Phenolate and carboxylate groups
4	Fire retardants	Rattan and areca particle board	Char forming agent
5	Ceramics	Pressed tiles	Increase the dry strength
6	Chelating agents	Agriculture, waste treatment, soil	Functional groups
7	Binders	Particle board, wall paper, coal briquettes, road	Calcium lignosulphonate
8	Emulsifiers	Linoleum flooring, wood based materials	Electrostatic & steric stabilization
9	Fillers	Fibre boards, ceramics, rubber, wood based materials, fodder pellets	Surface active characteristics
10	Dispersants	Construction, ceramics, coatings	Fine particles of lignosulphonates helps in distribution of fine particles
11	Batteries		Lead acid batteries
12	Carbon fibre	Aerospace, automotive, sporting goods	Highly efficient adsorption
13	Fertilizer	Agriculture sector	Slow release fertilizer
14	Adhesives	Wood industry	Magnesium lignosulphonate
15	Construction	• Concrete • Cement slurry • Plaster • Plaster blends	Plasticizer and water reducer, Filler

4 Conclusions

The demand and market of lignosulphonates shall grow due to its utililty in diversified end products including dust control, concrete additives, animal feed, binders, oil well additives and many more. The purity of lignosulphonates is crucial in its determination for its end use. The purified form of lignosulphonates has wider utility in comparison to crude spent sulphite liquor. Lignosulphonates are used as binders in wood-based materials, such as phenol-based adhesives, where they can serve as a partial substitute. Lignosulphonates have high molecular weight and contains various functional groups, by which they have found versatility in various end products. The lignosulphonates has already found its utility in various sectors on commercial scale and in the coming years, the market of lignosulphonates in the form of calcium lignosulphonates, sodium lignosulphonates, magnesium lignosulphonates shall flourish at the global scale.

References

1. Kazzaz AE, Feizi ZH, Fatehi P (2020) Technical lignin and its potential modification routes: a mini-review. Ind Crops Prod 154(9):112732
2. Mondal AK, Uddin MT, Sujan SMA, Tang Z, Alemu D, Begum HA, Li J, Huang F, Ni Y (2023) Preparation of lignin-based hydrogels, their properties and applications. Int J Biol Macromol 245:125580
3. Matsushita Y, Yasuda S (2005) Preparation and evaluation of lignosulfonates as a dispersant for gypsum paste from acid hydrolysis lignin. Bioresour Technol 96(4):465–470. https://doi.org/10.1016/j.biortech.2004.05.023
4. Gall DL, Ralph J, Donohue TJ, Noguera DR (2017) Biochemical transformation of lignin for deriving valued commodities from lignocellulose. Curr Opin Biotechnol 45:120–126
5. Tang X, Zhao K, Gao C, Gao X, Balan V, Wang W (2022) Isolation of lignin from anaerobically digested unhydrolyzed solids produced in a biorefinery. Agriculture 12(10). https://doi.org/10.3390/agriculture12101621
6. Vinardell MP, Mitjans M (2017) Lignins and their derivatives with beneficial effects on human health. Int J Mol Sci 18(6). https://doi.org/10.3390/ijms18061219
7. Bose S, Chakrabarty SK, Sarma US (2023) Extraction of lignosulfonate from jute sticks and its application in lead acid battery. Proc Natl Acad Sci 93:729–735
8. Areej F, Mohd Ashadie AK, Zakiah S, Ainun ZMA (2023) Pulping process for nonwoody plants (Chap 2). In: Ainun ZMA, Sapuan SM, Ilyas RA (eds) Pulping and papermaking of nonwood plant fibers. Academic Press, ISBN 9780323916257, https://doi.org/10.1016/B978-0-323-91625-7.00019-9
9. Alam MM, Greco A, Rajabimashhadi Z, Corcione CE (2024) Efficient and environmentally friendly techniques for extracting lignin from lignocellulose biomass and subsequent uses: a review. Clean Mater 13:100253
10. Kazzaz AE, Fatehi P (2019) Grafting strategies for hydroxy groups of lignin for producing materials. Green Chem 21(21), 154(9), 112732
11. Gonçalves S, Ferra J, Paiva N, Martins J, Carvalho LH, Magalhães FD (2021) Lignosulphonates as an alternative to non-renewable binders in wood-based materials. Polymers 2021, 13(23), 4196
12. Liu Q, Luo L, Zheng L (2018) Lignins: biosynthesis and biological functions in plants. Int J Mol Sci 19(2). https://doi.org/10.3390/ijms19020335

13. Ding X, Luo Z, Cheng F, Deng J (2021) Dust control performance enhancement of red sand via the synergistic application of Na- and Ca-lignosufonates. Int J Environ Sci Technol 19(8):7993–8006
14. Lorquin F, Lorquin J, Claeys-Bruno M, Rollet M, Robin M, Di Giorgio C, Piccerelle P (2021) Lignosulfonate is an efficient SPF booster: application to eco-friendly sunscreen formulations. Sustain Chem Pharm 24:100539. https://doi.org/10.1016/j.scp.2021.100539
15. Kiselev VI, Denisov VA, Korogodskii VG, Shnaider AV, Isupov YD (1989) Advantage of using lignosulfonates to lubricate ingot molds. Metallurgist 33, 238
16. Litters T, Liebenau A (2011) Lubricating greases containing lignosulfonate, the production thereof, and the use thereof. CA 2788157A1
17. Sidik MAM, Kadri A, Raj R, Gogoi BJ (2000) Use of sodium lignosulphonate in aqueous drilling fluid system for mud property enhancement. Int Res J Eng Technol (IRJET) 7(4):2229–2234
18. Browning WC, Chesser BG, Houston and Wood JL (1973) Oil based drilling fluid composition and process. Patent number 3738934
19. Aro T, Fatehi P (2017) Production and application of lignosulfonates and sulfonated lignin. ChemSusChem 10(9):1861–1877
20. Ugartondo V, Mitjans M, Vinardell MP (2008) Comparative antioxidant and cytotoxic effects of lignins from different sources. Bioresour Technol 99(14):6683–6687. https://doi.org/10.1016/j.biortech.2007.11.038
21. Chen M, Li Y, Liu H, Zhang D, Shi Q, Zhong X, Guo Y, Xie X (2023) High value valorization of lignin as environmental benign antimicrobial 18:100520
22. Huang C, Ma J, Zhang W, Huang G, Yong Q (2018) Preparation of lignosulfonates from biorefinery lignins by sulfomethylation and their application as a water reducer for concrete. Polymers 10(8): 841. https://doi.org/10.3390/polym10080841
23. Madyaratri EW, Iswanto AH, Nawawi DS, Lee SH, Fatriasari W (2022) Improvement of thermal behavior of rattan by lignosulphonate impregnation treatment. Forests 13(11). https://doi.org/10.3390/f13111773
24. Madyaratri EW, Ridho MR, Iswanto AH, Osvaldová LM, Lee SH, Antov P, Fatriasari W (2023) Effect of lignin or lignosulfonate addition on the fire resistance of areca (*Areca catechu*) particleboards bonded with ultra-low-emitting urea-formaldehyde resin. Fire 6(8):1–15. https://doi.org/10.3390/fire6080299
25. Lu W, Ye J, Zhu L, Jin Z, Matsumoto Y (2021) Intumescent flame retardant mechanism of lignosulfonate as a char forming agent in rigid polyurethane foam. Polymers 13(10):1–13. https://doi.org/10.3390/polym13101585
26. Lu W, Li Q, Zhang Y, Yu H, Hirose S, Hatakeyama H, Matsumoto Y, Jin Z (2018) Lignosulfonate/APP IFR and its flame retardancy in lignosulfonate-based rigid polyurethane foams. J Wood Sci 64(3):287–293. https://doi.org/10.1007/s10086-018-1701-4
27. Ruwoldt J (2020) A critical review of the physicochemical properties of lignosulfonates: chemical structure and behavior in aqueous solution, at surfaces and interfaces. Surfaces 3(4):622–648. https://doi.org/10.3390/surfaces3040042
28. Breilly D, Fadlallah S, Froidevaux V, Colas A, Allais F (2021) Origin and industrial applications of lignosulfonates with a focus on their use as superplasticizers in concrete. Constr Build Mater 301, 124065, https://doi.org/10.1016/j.conbuildmat.2021.124065
29. Gulcin İ, Alwasel SH (2022) Metal ions, metal chelators and metal chelating assay as antioxidant method. Processes 10(1):132. https://doi.org/10.3390/pr10010132
30. Kurup GG, Adhikari B, Zisu B (2019) Recovery of proteins and lipids from dairy waste water using food grade sodium lignosulphonate. Water Resour Ind 22:100114
31. Malode KS, Pandey IK, Bhonde S, Kothwade P (2023) Benefits of novel organic chelated lignosulfonate fertilizers for the plant, soil and environment: a brief review. Int J Curr Sci 13(2):156–159
32. Agrawal A, Kaushik N, Biswas S (2014) Derivatives and applications of lignin – an insight. Scitech J 01(07):30–36

33. Deng W, Feng Y, Fu J, Guo H, Guo Y, Han B, Jiang Z, Kong L, Li C, Liu H, Nguyen PTT, Ren P, Wang F, Wang S, Wang Y, Wang Y, Wong SS, Yan K, Yan N, Yang X, Yuanbao Z, Zhang Z, Zeng X, Zhou H (2023) Catalytic conversion of lignocellulosic biomass into chemicals and fuels. Green Energy Environ 8(1):10–114
34. Vinod A, Pulikkalparambil H, Jagadeesh P, Rangappa SM, Siengchin S (2023) Recent advancements in lignocellulose biomass-based carbon fiber: synthesis, properties, and applications. Heliyon 9(3). https://doi.org/10.1016/j.heliyon.2023.e13614
35. Ruwoldt J, Planque J, Øye G (2020) Lignosulfonate salt tolerance and the effect on emulsion stability. ACS Omega 5(25):15007–15015. https://doi.org/10.1021/acsomega.0c00616
36. Czaikoski A, Gomes A, Kaufmann KC, Liszbinski RB, de Jesus MB, da Cunha RL (2020) Lignin derivatives stabilizing oil-in-water emulsions: technological aspects, interfacial rheology and cytotoxicity. Ind Crops Prod 154, 112762
37. Ruwoldt J (2023) Emulsion stabilization with lignosulphonates. In: Sand A, Tuteja J (eds) Lignin - Chemistry, Structure, and Application, IntechOpen, doi: 10.5772/intechopen.107336
38. Kruželák J, Hložeková K, Kvasničáková A, Džuganová M, Chodák I, Hudec I (2023) Application of plasticizer glycerol in lignosulfonate-filled rubber compounds based on SBR and NBR. Materials 16(2). https://doi.org/10.3390/ma16020635
39. Sappi (2016) Dust suppressant and surface stabiliser (Lignex). Sappi. Retrieved [February, 2026], from https://www.sappi.com/en-lc/products-services/products/dust-control-and-road-stabilisation
40. Magina S, Barros-Timmons A, Evtuguin DV (2021) Synthesis of lignosulfonate-based dispersants for application in concrete formulations. Materials 14(23). https://doi.org/10.3390/ma14237388
41. Xu C, Ferdosian F (2017) Utilization of lignosulfonate as dispersants or surfactants. In Conversion of Lignin into Bio-Based Chemicals and Materials (pp. 81–100). Berlin, Heidelberg: Springer. https://link.springer.com/chapter/10.1007/978-3-662-54959-9_5
42. Niaounakis M (2015) Building and construction applications. In: Biopolymers: applications and trends. https://doi.org/10.1016/b978-0-323-35399-1.00010-7
43. Areskogh D (2011) Structural modifications of lignosulphonates (Doctoral dissertation). KTH Chemical Science and Engineering, TRITA-CHE Report 2011:26, Royal Institute of Technology, Stockholm, Sweden
44. Gond R, Asfaw HD, Hosseinaei O, Edström K, Younesi R, Naylor AJ (2021) A lignosulfonate binder for hard carbon anodes in sodium-ion batteries: a comparative study. ACS Sustain Chem Eng 9(37):12708–12717
45. Naylor A (2020) Lignosulfonate binders for green next-generation battery electrodes. Project report (final). CMC, pp 1–14
46. Zhou W, Chen M, Tian Q, Chen J, Xu X, Han X, Xu J (2021) Stabilizing zinc deposition with sodium lignosulfonate as an electrolyte additive to improve the life span of aqueous zinc-ion batteries. J Colloid Interface Sci 601:486–494
47. Bayaguud A, Luo X, Fu Y, Zhu C (2020) Cationic surfactant type electrolyte additives enables three dimensional dendrites free zinc electrodes for stable zinc ion batteries. ACS Energy Lett 5(9), 3012–3020
48. Yhobu Z, Siddiqa A, Padaki M, Budagumpi S, Nagaruja DH (2022) Lignocellulose biopolymers and electronically conducting polymers: towards sustainable energy applications. Energy Fuels 36(24):14625–14656
49. Simmons T, Mukherjee R (2015) WO2015057652 - High performance cathode materials for lithium sulfur batteries from lignosulfonate. In: WIPO, Publication Number WO/2015/057652, Retrieved [February, 2026], https://patentscope.wipo.int/search/en/WO2015057652
50. Lu H, Cornell A, Alvarado F, Behm M, Leijonmarck S, Li J, Tomani P, Lindbergh G (2016) Lignin as a binder material for eco-friendly Li-ion batteries. Materials 9(3):1–17
51. Ahankari SS, Mohanty AK, Misra M (2023) Nanomaterials from renewable sources for emerging applications. In: Emerging materials and technologies, technology & engineering. CRC Press, 1–542

52. Rajlakshmi, Dutta S, Saravanabhupathy S, Jayalseelan P, Kailasam AA, Banerjee R (2024) Jute stick-based industrial artefacts towards sustainable development goals. Ind Crops Prod 215:118599
53. Myrvold B (2005) Interactions between lignosulphonates and the components of the lead acid battery. Part 2. Interactions with the electrolyte and lead sulphate growth. J Appl Electrochem 35(6):573–579. https://doi.org/10.1007/s10800-005-1821-6
54. Chang FC, Yen SH, Wang SH (2018) Developing lignosulfonate-based activated carbon fibers. Materials 11(10):1–12
55. Mainka H, Täger O, Körner E, Hilfert L, Busse S, Edelmann FT, Herrmann AS (2015) Lignin - an alternative precursor for sustainable and cost-effective automotive carbon fiber. J Mater Res Technol 4(3):283–296
56. Wang S, Bai J, Tendo M, Wang Q, Xiang H, Tang J, Zhu M (2022) Lignin-based carbon fibers: formation, modification and potential applications. ScienceDirect 7:578–605. www.transparentmarketresearch.comI.
57. Bengtsson A, Hecht P, Sommertune J, Ek M, Sedin M, Sjöholm E (2020) Carbon fibers from lignin-cellulose precursors: effect of carbonization conditions. ACS Sustain Chem Eng 8(17):6826–6833
58. Razzaq MA, Rahman A, Mohammed ABA (2024) Lignin based carbon fiber toward sustainability: opportunities and challenges. Elsevier B.V.
59. Luo Y, Xia Y, Wang C, Chen J, Ou L (2023) Application of calcium lignosulphonate as an environment friendly depressant in the Cu-As separation by froth flotation at low alkalinity. J Clean Prod 406, 137073, https://doi.org/10.1016/j.jclepro.2023.137073
60. Tapley B, Yan D (2003) The selective flotation of arsenopyrite from pyrite. Minerals Engineering, 16(11):1217–1220. https://doi.org/10.1016/j.mineng.2003.07.017
61. Clarke N, Peng D, Clarke JL (2023). Innovation for environmentally-friendly food production & food safety in china. In: Sustainable agriculture and food security. Springer Science and Business Media LLC, pp 1–252
62. Raji, Adeola B (2024) Chemical composition and bioactivities of African spinach (*Amaranthus Virdis*) grown on four different sites, Kwara State University, Nigeria
63. Meier JN, Fyles JW, Mackenzie AF, O'Halloran IP (1993) Effects of lignosulfonate-fertilizer applications on soil respiration and nitrogen dynamics. Can J Soil Sci 73(2):233–242
64. Ijaz N, Dai F, Meng L, Rehman ZU, Zhang H (2020) Integrating lignosulphonate and hydrated lime for the amelioration of expansive soil: a sustainable waste solution 254:119985
65. Karthouser J, Biziks V, Moi C, Militz H (2021) Lignin and lignin derived compounds for wood application 26(9):2533
66. Balea G, Lunguleasa A, Zeleniuc O, Coşereanu C (2022) Three adhesive recipes based on magnesium lignosulfonate, used to manufacture particleboards with low formaldehyde emissions and good mechanical properties. Forests 13(5), 737
67. Gonçalves S, Martins J, Paiva NT, Paiva D, Carvalho LH, Magalhães FD (2023) The potential of visible spectroscopy as a tool for the in-line monitoring of lignin methylolation. Polymers 15(1), 178
68. Bartolome MJ, Padhi SSP, Fichtberger OG, Schwaiger N, Seidl B, Kozich M, Nyanhongo GS, Guebitz GM (2022) Improving properties of starch-based adhesives with carboxylic acids and enzymatically polymerized lignosulfonates. Int J Mol Sci 23(21):13547
69. Dunky M, Mittal KL (2023) Biobased adhesives: sources, characteristics, and applications, adhesion and adhesives: fundamental and applied aspects. Wiley, pp 1–768
70. Gonçalves S, Paiva NT, Martins J, Paiva D, Magalhães FD, Carvalho LH (2024) Effect of lignosulphonates on the moisture resistance of phenol-formaldehyde resins for exterior plywood. Materials 17:3715
71. Amulya G, Moghal AAB, Basha BM, Almajed A (2022) Coupled effect of granite sand and calcium lignosulphonate on the strength behavior of cohesive soil. Buildings 12(10):1687
72. Wu D, She W, Wei L, Zuo W, Hu X, Hong J, Miao C (2020) Stabilization mechanism of calcium lignosulphonate used in expansion sensitive soil. Adv Mater 35:847–855

73. Wang J, Wang J, Qin X, Wang Y, You Z, Liu H, Shao M (2020) Superfine MnO nanowires with rich defects toward boosted zinc ion storage performance. ACS Appl Mater Interfaces 12(31):34949–34958
74. Bampidis V, Azimonti G, de Lourdes Bastos M, Christensen H, Dusemund B, Kouba M, Kos Durjava M, López-Alonso M, López Puente S, Marcon F, Mayo B, Pechová A, Petkova M, Ramos F, Sanz Y, Villa RE, Woutersen R, Aquilina G, Bories G, Gropp J, Nebbia C, Finizio A, Focks A, Teodorovic I, Innocenti ML, Tarrés-Call J (2020) Safety of lignosulphonate for all animal species. EFSA J 18(2) doi: 10.2903/j.efsa.2020.6000
75. Fazeli M, Mukherjee S, Baniasadi H, Abidnejad R, Mujtaba M, Lipponen J, Seppälä J, Rojas OJ (2024) Lignin behind the status quo: recent and emerging composite applications. Green Chem, 26, 593–630
76. Mu L, Wu J, Matsakas L, Chen M (2018) Lignin from hardwood and softwood biomass as a lubricating additive to ethylene glycol. Molecules 23(3), 537
77. Zimáková J, Fryda D, Vaculík S, Bača P, Bouška M (2018) Examination of impact of lignosulfonates added to the negative active mass of a lead–acid battery electrode. J Energy Storage 18:229–238

Recent Advances in Chemical Recovery Furnace for the Efficient Combustion of Black Liquor

Jaykumar Bhasarkar, Suraj Gawade, and Prasad Patil

Abstract The main aim of this chapter is to give insight into the recovery furnace. In recent times, the recovery furnace is not only used to recover the cooking chemicals in the paper industry but also generate the energy in terms of electricity. In modern days, the demands of the electricity consumptions increase and therefore in order to cater these demand, the new modern recovery furnace need to be developed. In this chapter, we have discussed the early recovery furnace and their development in various functions. The corrosion and erosion is the major issue in any recovery furnace as the concentrated black liquor is the feedstock. Black liquor contains several organic, inorganic and non process elements and hence it is very essential to control the corrosion in the furnace by increasing the heat transfer rate. Several possible solutions have been discussed to enhance the capacity of the furnace by controlling the corrosion and erosion problems incorporated in the recovery furnace. We believe that this chapter gives more insight into the recovery furnace which further helps in the development of this research area.

Keywords Recovery furnace · Drying · Devolatalization · Char combustion · Smelt

1 Introduction

Recovery furnaces are important key in several industries. The recovery technology is very older technology than the recovery furnace. Recovery technology was not invented till the pulping technology was introduced in the industry. However, the

---

J. Bhasarkar (✉) · P. Patil
Deparmant of Chemical Engineering, Laxminarayan Innovation Technological University, Nagpur, Maharashtra, India
e-mail: jaykumar.ppt@gmail.com

S. Gawade
Deparment of Pulp and Paper Technology, Laxminarayan Innovation Technological University, Nagpur, Maharashtra, India

S. Sundaramurthy et al. (eds.), *Black Liquor Waste Management in the Pulp and Paper Industry*, Sustainable Materials and Technology,
https://doi.org/10.1007/978-981-96-4237-3_6

recovery furnace has been built with some features that is compatible with the old technology. The use of recovery boiler/furnace was not limited for the paper industries but it has wide application in several industries such as, petroleum, fertilizer, mining, thermal power plant, etc. Early days, the chemical furnace mainly focused on the recovery of chemicals which ultimately improved the profit of the pulp and paper manufacture industry. Later, it was noticed that, the recovery furnace can also be used for in-situ electricity generation. During chemical recovery with the furnace it is useful to enhance the smelt, i.e. un-burnt inorganic compounds recovery. Pulp and paper industry is the major industry that uses the recovery furnace for the recovery of spent cooking liquor. Generally, for pulping process, Kraft method has been adopted (Kraft pulping process) [1]. In order to recover the Kraft pulping chemicals, Kraft recovery furnace is used. Kraft recovery furnace is the largest process to generate the steam from the woody raw material or biomass [2, 3]. The basic mechanism in the recovery furnace is to burn the organic material and recover the unburnt material such as inorganic compounds as smelt. In the modern paper industry, the energy (steam and electricity) is generated with the help of the recovery furnace and this produced energy is sufficient to utilize for the different processes involved in the paper mills [4]. In Kraft process, basically the recovery section is divided in two sections, first chemical recovery and fiberline which is related to pulp. Generally, the black liquor generated from the washer section pass through the evaporator where the diluted black liquor was concentrated. The concentrated black liquor mainly consists of organic and inorganic compounds. In the recovery boiler, combustion of concentrated black liquor is carried out to produce sodium sulfide and sodium carbonate [5, 6]. Sodium carbonate is further causticize in order to get sodium hydroxide [7]. Several reaction mechanisms takes place in the recovery boiler. Combination of pulping, washing and bleaching process in integrated pulp and paper industry is allowed to maintain the Kappa number as low as possible. This lower Kappa number signifies the increased organic content in the black liquor [8]. The properties and constituents present in the black liquor generated from the washing of pulp section is depicted in Table 1 [7].

1.1 Function of the Recovery Furnace

The primary function of the recovery boiler is to burn the organic content present in the black liquor. The burnt material generates heat which ultimately leads to produce the high pressure steam and thereby the electricity. In addition, the major role of the recovery furnace is to recover chemicals from the unburnt inorganic material as discussed earlier. However, sulfur content of ~4–5% is present in the black liquor which is released during the combustion process and therefore it may hamper the surrounding environment. Optimization of process parameter is the main focus in order to reduce the sulfur and nitrogen compounds (SOx, NOx) to the environment [9, 10]. As discussed in the previous section, concentrated black liquor from the evaporator is fired to the recovery furnace though nozzles in the form of droplets. The black liquor droplets are entered in the furnace through liquor guns and have

Table 1 Properties of black liquor [7]

Property	1982	1992	2002
Liquor dry solids kg dry solids/t pulp	1700	1680	1760
Sulfidity, $Na_2S/(Na_2S + NaOH)$	42	45	41
Black liquor HHV, MJ/Kg dry solids	15.0	13.9	13.0
Liquor dry solids, % Elemental analysis, %weight	64	72	80
C	36.4	34	31.6
H	3.75	3.5	3.4
N	0.1	0.1	0.1
Na	18	18.4	29.8
S	5.4	5.9	6
Cl	0.2	0.4	0.8
K	0.75	1.0	1.8
Cl/(Na + K), mol %	0.70	1.37	2.49
K/(Na + K), mol %	2.39	3.10	5.07
Net heat to furnace, kW/kg dry solids	13,600	12,250	11,200
Combustion air required; m^3n/kg dry solids	4.1	3.7	3.4
Flue gas produced; m^3n/kg dry solids	4.9	4.3	3.9

two major methods namely, wall firing and suspension firing. Several reaction mechanisms takes place in the recovery boiler. The black liquor droplets undergo a series of reactions such as combustion, gasification and drying [11]. After homogeneous combustion of black liquor, solid unburnt residue is still present in the recovery furnace and therefore it forms a char bed at the bottom of the furnace [12].

1.2 Development in Recovery Furnace Design

Major modifications in recovery furnace have been performed due to several reasons such as investment cost, scale up and energy demand. These are the major factors that directly affect the performance of the recovery boiler and therefore the changes in design. The size of the furnace has increased day by day to handle more material or chemicals in recent times. At the beginning of the 1980s, the capacity of the furnace was 1700 tons/dry solids/day. Later the capacity of the furnace was enhanced to ~3000–3500 t of dry solid/day in the 1990s [2]. The capacity of the recovery furnace has increased due to low water content in black liquor and uniform liquor spray [13]. As discussed earlier, the black liquor was treated at the evaporator to increase the dry solid concentration and therefore it increases the steam generation. However, the main problem is associated with the handling of high dry solid content due to their high viscosity. Liquor heat treatment is the method to reduce the viscosity of the liquor. During pulping and washing, several small particles such as organic chemicals are incorporated in the system which later acts as dry solids. The ratio of dry black liquor to wet black liquor (*w*/*w*) is defined as dry solids [14]. The calorific value of the black liquor is an important property and it is affected by the moisture content present in the liquor [15]. If there is moisture content in the liquor it reduces the heating value of the liquor.

1.3 Recovery Boiler Steam Pressure and Temperature

Several modifications have been done in the recovery furnace from 1900s to till date [2]. However, the development in steam pressure and temperature was not taking much more time. McCarthy has investigated the maximum pressure achieve was 10 MPa at ~480–500 °C. But this pressure and temperature range was reduced due to some safety purpose [14]. Later several recovery boilers have been installed and all were operated at a pressure of 8.0 MPa.

1.4 Safety in the Recovery Furnace

Safety is the major concern in the recovery furnace as several harmful reactions take place in the furnace. If small drop of water or moisture reacts with the black liquor which contain high amount of dry solids then the reaction takes place is exothermic in nature. Due to the exothermic nature of the reaction, explosion might occur at the smelt-water system. The explosion is a natural physical reaction when the water molecules get evaporates faster when it comes in contact with the hot smelt. This phenomenon of evaporation leads to an enhancement in the volume, temperature and pressure in the system. The wall of the recovery furnace is built to sustain the limited pressure range and therefore the sudden increment in the pressure causes to rupture a surface or bend the surface. Water contamination is the primary focus and it is assured that the water content should not enter in the recovery furnace. If any water molecules have entered in the furnace, then sudden shutdown take place in order to avoid any consequences.

2 Chemical Mechanism in the Recovery Furnace

In the chemical recovery furnace, several stages of reactions takes place. The main aim of this mechanism is combustion of black liquor. The combustion of black liquor is mainly dependent on the type of raw material used and dry solid content present in the liquor. As mentioned in the previous section, the organic materials are burnt and the inorganic material form a smelt by reduction reaction. Generally, black liquor

Table 2 Stages involved during black liquor combustion in the recovery furnace

Stage	Characterized by	Time scale in furnace for 2 mm droplet
Drying	Water evaporation Constant diameter after initial swelling	0.1–0.2 s
Devolatilization	Appearance of flame, ignition Swelling of droplet Release of volatiles	0.2–0.3 s
Char burning	Disappearance of flames Decrease in diameter Reduction reaction	0.5–1.0 s
Smelt	Constant or increasing diameter Reoxidation	Long

has low heating value due to the presence of more inorganic contents. Therefore, it is very difficult to perform the combustion reaction in reality [16]. The black liquor is sprayed in the recovery furnace in the form of droplets by the liquor guns. Several liquor guns are used to spray the droplets. The size of the droplets should be maintained as small as possible. The smaller sizes of the droplets are helpful to enhance the combustion process and also increase the temperature. The average size of the liquor droplets can vary from 2 to 3 mm. Chemical reaction in the recovery boiler is eco-friendly and does not generate any harmful pollutant from the furnace [17, 18]. Several researches have been performed to optimize the system to enhance the efficiency of the process. But due to complexity in the reaction, it is very difficult to control the chemical reactions in the furnace. Generally, several simultaneous reactions are carried out in the different reaction zones in the recovery furnace [19].

The chemical reaction mechanism in the recovery furnace consists of four stages as depicted in Table 2. During the combustion of black liquor, the droplets of black liquor undergo the swelling mechanism. Basically, the swelling of droplets are important phenomena for any process where thermal conversion of black liquor takes place. There are different stages involved in the swelling of droplets and the combustion of the black liquor [12, 20]. The detail discussion of all four stages of black liquor combustion is as follows.

2.1 *Drying*

Generally, drying is the important process that is carried out in the furnace. Evaporation of moisture or water content from the black liquor takes place in order to avoid any unwanted reaction or accident. Energy in terms of heat is the most essential part to evaporate the water. In the recovery furnace, the heat transfer to the droplets is not sufficient and therefore the heat flux is also limited to evaporate the maximum amount of water. In the drying stage, the liquor droplet starts swelling and after some time, the diameter of the droplets are constant [7]. Thc black liquor droplets size increased by ~1.4 times to its initial diameter. As the water molecules are evaporated, it affects the physical properties of the black liquor such as density of the liquor decreases. Even after the density decreases the diameter of the droplets is maintained constant throughout the stage. Initially, the drying rate is slow and it does not completely remove the water content. At the end of this stage, maximum ~5% moisture content is present in the black liquor [7, 21]. Some studies have reported that the drying rate is kept constant as different dry solid contents are used at 700–800 °C with some specific raw material such as pine or birch [7, 22, 23].

2.2 Devolatilization of Black Liquor

In this stage, the dry solid concentration in black liquor increases due to the removal of maximum water content in the drying zone. Temperature plays an important role in the devolatilization stage [17]. In this stage the temperature and dry solid content increases and therefore the rate of reaction increases by means of lower activation energy of the reaction. During the devolatilization process, the organic material is burnt and lower molecular weight compounds are released from the furnace [24]. The size of the droplets increases due to the removal of volatile compounds from the black liquor and it appears as a flame. The total duration of this stage varies from 2 to 3 s as depicted in Table 2. During the process the droplets swell considerably. This swelling process is persistent till the devolatilization process is not completed. Generally, gasification or pyrolysis has been adopted for the devolatalization of black liquor [25]. For the efficient devolatalization of black liquor, two main conditions should be satisfied (a) there must be gas generation and (b) black liquor droplets must have plasticity properties. After swelling of black liquor particles it should be more porous in nature. The swelling of particles mainly depends on the type of raw material used such as softwood or hardwood, cooking time and lignin to aliphatic acids ratio. During the devolatilization of black liquor several gasses are released as discussed earlier. Sulfur is the element which is released during this process in the form of dimethyl sulfide or methyl mercaptant. Another gas, i.e. hydrogen sulfide is also released during the process after the gasses are evacuated from the droplet [26].

2.3 Char Combustion

This reaction starts when the devolatilization of black liquor process in the verge of completion. The indication of completion of the devolatilization process is when the visible flame vanishes. In general, the char combustion and devolatilization process overlap each other as one process is close to end, while another process is at the beginning. During the combustion, all inorganic materials are converted into the carbon char. The produced char mainly consist of four compounds such as carbon, sodium sulfate, carbonate and sulfide. The elemental analysis of char revealed that the most portion are covered by carbon (~2/3rd) and the remaining portion is the hydrogen without any traces of oxygen. The reduction rate of char is ~50–55% with continuous burning as the temperature increases in the furnace by an external source. During the reduction reaction, carbon plays an important role to form sodium sulfide. Therefore, the rate of reaction mainly depends on the carbon content available in the char [25].

2.4 Smelt Reaction

After completion of char combustion, the inorganic materials remain in the furnace. As the droplets of the liquor initially swell, later the droplets gets shrunk. In this process, oxygen molecules react with the smelt and it is converted into sodium carbonate by reoxidation reaction. During the smelt reaction, the oxidation of smelt is also converted to form sodium sulfate. Later these compounds get reconverted to form the cooking chemicals.

3 Evaluation of Recovery Furnace Design

Several changes have been made in the Kraft recovery furnace to improve the performance of the furnace. Pulping process is the main key factor to affect the performance of the boiler. Any changes done in the pulping process it changes the properties of the black liquor and therefore it changes the performance of the recovery furnace. This section provides insight into the development in the recovery furnace.

3.1 Two Drum Recovery Boiler

Generally, the two drum design of recovery boiler is mostly used in the industry. The generation of steam is about to 85 bar pressure and temperature of the steam is 480–500 °C. This type of boiler has the capacity to handle the solid content of 1700–2000 t of dry solids/day. Two drum recovery furnaces mainly consist of the firing unit which is stationary and 3 levels for air intake, i.e. primary, secondary and tertiary. This recovery furnace is an efficient boiler to reduce the emission of sulfur compounds.

Two drum recovery furnaces consist of a water screen which covers the superheaters from the corrosion and irradiation. As the temperature of the furnace reduces, which causes low flue gas temperature and therefore it decreases the material deposited on the surface of the superheater. In this type of recovery furnace, only vertical type of economizer has been used and it was a development in the recovery boiler by replacing the horizontal economizer. The main advantages of the vertical economizer are to reduce the fouling. Recently, the two drum boiler are getting modified and they are converted to a single drum recovery furnace.

3.2 *Modern Recovery Furnace*

As discussed in the previous section, the two drum recovery furnace is modified and converted into a single drum recovery furnace. Single drum recovery furnace has advantages that the vertical economizer is being used to increase the resistance to fouling. The use of vertical economizer is very simple to clean the tubes and remove the deposits from the surface. Another advantage is that the spacing between the tubes can be increased upto 400 mm. This large spacing is provided to remove the deposits easily and therefore it can reduce the fouling. Several modifications have been made to reduce the fouling and lower the corrosion. The operating parameters of a single drum recovery furnace mainly depends on the dry solid concentrations. The black liquor dry solid concentration increases it is required to increase the flow rate. Increasing the flow rate will lead to reduce the heat loss from the flue gases. The major advancement and modification has been done in single drum recovery furnace design. This modification in recovery furnace is mostly affected by the water quality. Another advantage of single drum recovery furnace is easy to use and good safety compared to the two drum system. Single drum system generates higher steam pressure with higher capacity. Austenitic material is used to make a lower portion of the superheater. This material helps to reduce the corrosion. Temperature of the surface decrease as the velocity of the steam is higher. Again the lower temperature of the tube surface is useful to reduce the corrosion. For the uniform flow of steam in the tubes, it is ensured by maintaining the high steam side pressure.

Air system

Fuels are made up of both organic and inorganic substances that undergo reactions in the furnace. For combustion to continue, air is required. The primary needs for the boiler air system include achieving maximum mixing and maintaining an optimal air ratio profile. The delivery of air in a boiler system typically occurs at multiple horizontal elevations to achieve thorough combustion and minimize emissions [13]. It has an inlet duct with a silencer, venturi for air flow measurement, air blower, air heater, distribution ducts, and air ducts. Air blowers and dampers in the apertures regulate the air flow. The adjustment of the rotating speed of the blower is managed by dampers in the duct and to control airflow through the blowers inlet vane dampers are required. The desired duct air pressures are produced as desired by the air flow control with dampers in the openings. Although high duct pressure uses more blower power but it guarantees consistent flow across every outlet. There is normally high air intake in the boiler house. By doing this, it makes sure that the boilers walls and openings lose heat and have an even vertical temperature profile. Air blower power consumption is lowered when the air blower is placed upstream from the air preheaters [27]. Following the air heater, the air is divided into distinct ducts to ensure even distribution to various sides of the boiler. It is crucial to employ modest design velocities since even minor form changes in parallel air ducts can result in significant flow variations [27, 28]. Different kinds of air heaters are employed to regulate the air temperature. In cases where the flue gas is free of impurities, it can be utilized to preheat the air. Cross flow air heaters with multiple pass steel tubes are commonly used in bark boilers. Recovery boilers as well as steam and feed water air heaters are utilized in contaminated environments. An air heater's function is to raise the furnace temperature. This raises the temperature of steam and reduces the required heat transfer surfaces.

4 Alternatives in Recovery Furnace Design

There are several options available while modifying the recovery furnace. In recovery furnace mainly modifications have been made in the design of the superheater, material alternatives for the tubes and the position of the boiler bank or economizer. Air level type is another area for modification in recovery furnace.

4.1 Design Specification

Capacity of the recovery furnace is the main specification usually suggested by the industry as per the requirement. While designing the recovery furnace, information on several parameters is required to get the appropriate design. These parameters such as dry solid content, heating value of the black liquor and elemental analysis of the black liquor are known. Evaporation operation is another important process that is very helpful to design the furnace. The size of the furnace depends on the dry solid content in the black liquor. The elemental analysis of the black liquor assists to calculate the heat released in the recovery furnace. The maximum continuous steam rating for the steam flow is calculated by the water and steam value which is used in the furnace. Generally, the ash containing black liquor is fed to the recovery furnace and therefore the characteristics of the ash free black liquor and ash-black liquor are different. So, it is very necessary to design the recovery furnace by considering both characteristics of the black liquor.

4.2 Modification in the Air System

To reduce the recalcitrant emission from the recovery furnace it is necessary to focus on the air level system which is in the furnace. The design of the air system should be proper so that it can achieve the solid operations in the furnace. The development of the air system is a continuous process as long as the recovery furnace exists. The targeted value for the air system in the furnace has changed regularly. Initially in 1940s, the recovery furnace had only two level of air system, i.e. primary and secondary. The purpose of the primary air system was to complete the reduction reaction in the furnace. However, the use of secondary air system was for the oxidation reaction. Later, the development has been made in the recovery furnace and the secondary air system is installed at 2 m below the liquor guns. In 1954s, three levels of air system has been used i.e. primary, secondary, and tertiary. In latest development, the recovery furnace has consisted of multilevel air system. The multilevel air system reduces the sulfur emission to the environment from the recovery furnace.

5 Enhancement in Heat Transfer

When the recovery furnaces are designed, one of the major factors that need to be considered is heat transfer in the recovery furnace. A heat transfer phenomenon is the important and essential parameter in the furnace. However, the heat transfer in the furnace mainly depend on the type of material used. During the process, corrosion is the major issue and the main focus is to reduce the corrosion in the furnace. The corrosion is differentiated into several types based on the water quality, and temperature range. Water type corrosion is based on the water and steam quality, temperature type corrosion depends on the high temperature or low temperature. High temperature corrosion is observed in the superheater tubes; however, the low temperature corrosion occurs at the economizer.

5.1 Design of Furnace and the Material

The design of a furnace deals with the identification of the most suitable material which provides better results. Basically, the material used for the base surface is very much important in terms of safety as the smelt is generated on the base surface of the furnace. Most of the times the leakages can be observed in the lower portion of the furnace and these leakages may be the very dangerous. Therefore, the significant protection is to be provided for the lower surface of the recovery furnace. The selection of the correct material is the challenging part by considering the mechanical strength, corrosion protection and economy of the process. In modern recovery furnace, the membrane type material is widely used for the furnace wall. In addition, alloy material is another potential material to protect the corrosion. Welded material is also being used for the wall construction due to its lower cost. During the pyrolysis process, it is essential to protect the tubes from the high temperature. To maintain the temperature range, water circulation is the important process. Therefore, the design of the water circulation unit is also important which further maintains

Table 3 List of materials used for the furnace tube [2]

	Carbon steel	304 L	Alloy 825	Alloy 625
Main element	Fe	20Cr-10Ni	20Cr-40Ni	20Cr-60Ni
Thermal expansion (10^{-6})/°C	13.5	17.5	14.9	13.9
Thermal conductivity W/m.°C	41	19	16	14
Corrosion resistance	Low	Moderate	Excellent	High

the floor temperature. During this process, the flow velocity is kept at 0.5 m/s for the better control of the temperature. Another parameter that needs to be considered for the better performance is the floor angle. The floor angle is kept upward in all the modern furnaces. This section provides the detail information about the design of different section involved in the recovery furnace. As discussed earlier, material selection is the prominent criteria for designing a recovery furnace. Generally, the selection of material is based on the water quality, air level systems, corrosion and water erosion take place in the furnace. Superheater, boiler bank and economizer are the sections where most chances of corrosion that takes place and here, we have discussed the material and design of these sections.

5.2 *Recovery Furnace Tube Material*

Several different type of material has been used for the furnace tube listed in Table 3. Several literatures have been reported that new materials are used for the furnace tube and older material has to be replaced with new one. Initially carbon tube was a potential material used as a tubing material before compound material was introduced in the furnace tube application. However, the above portion of the air level is always made using carbon steel which can resist corrosion issues.

Makipaa and Backman have studied the different materials for the corrosion resistance in polysulfide solution. They have reported that, composite material of Sanicro 38 was found to be the best corrosion resistant among all the steel material. The test results revealed that this alloyed composite material is corrosion resistant but it was very expensive [29].

5.3 *Superheater Design*

Corrosion in the superheater of the recovery furnace is the major issue. Recovery furnace has significantly more issue with superheater corrosion compared to coal fired boiler. Several inorganic and alkali particles such as chlorides come in contact with the superheater in the form of gas and therefore it leads to enhance the corrosion on the superheater. Due to alkali or chloride on the surface and sulfidation mechanism

Table 4 Energy available for superheaters [7]

Liquid dry solid, %	60	65	70	75	80	85	LHT
Flue gas flow, kg/s	221.0	216.9	212.8	209.5	206.7	204.2	197.2
Furnace exit temp., °C	913	926	936	943	949	953	937
Cp gas, kj/kg. °C	1.391	1.372	1.355	1.340	1.324	1.314	1.321
Steam flow, kg/s	122.8	26.6	129.9	132.8	135.3	137.5	132.8
Heat for superheater, MW	109.9	110.4	109.7	108.8	107.5	106.4	98.6
Heat for superheater, kW/kg	894.6	871.3	844.0	819.2	794.3	773.6	742.7

reduces the tube thickness of the superheater. Tube surface resist more corrosion if the chromium content is more in the tube material [30]. For selecting the superheater material, steam outlet temperature is the important parameter. It is very well known that the steam temperature should be kept lower than the melting temperature of the deposits in order to ease removal of the deposits. At the end of the process, the corrosion rate is increased due to higher temperature range in the superheater. Many literatures have suggested the suitable temperature rate to reduce the corrosion rate in the superheater. Thumb rule says that the enhancement in corrosion rate by increasing the tube temperature by 10 °C. Superheater position in the furnace also affects the temperature range and therefore it changes the corrosion rate. Position of superheater can be placed in such a way that it can avoid direct contact with the furnace radiations. The heat generated and avaibility for the superheater is depicted in Table 4. The result shows that, when the dry solid content increases in the black liquor, it reduces the flue gas temperature and increases the furnace exit temperature.

As discussed in the previous section, the chromium content material can reduce the corrosion rate in the superheater. Carbon steel is the potential material used for the primary superheater due to direct exposure of the furnace radiation. Carbon steel can be used as a primary material for the superheater; however, 1–3% chromium is used for the secondary and tertiary superheater. Fujisaki et al. [4] have investigated that, when the chromium content increases in the tube material, the corrosion rate decreases significantly.

5.4 Economizer Design

Initially, the recovery furnace has horizontal type economizer. Horizontal economizers have several issues regarding the plugging problems. To overcome this issue the new design of economizer was introduced in a cross flow manner. The result of this type of economizer revealed that the heat transfer coefficient was found to be low and the plugging problem was still unresolved. Later, the economizer was further modified and vertical type of economizer was used in the recovery furnace. In the economizer, the reduction in tube thickness is observed due to corrosion erosion due to water, and the sulfidation process. The economizer has several weld joint and each joint creates various problems in the recovery furnace. In order to reduce these problems associated with the weld join, it is necessary to reduce the weld joint as much as possible in the economizer. Therefore, the material used in the construction of an economizser should be continuous without any butt joints.

6 Future Recovery Boiler

In recent times, the higher steam pressure requirement is the major issue. The application of the chemical recovery furnace is not limited to recover the cooking chemicals in the pulp and paper industry but also to generate the electricity. The demand of electricity generation increases by 7% due to enhancement in the steam pressure and temperature. The enhancement in the process condition results to generate extra ~ 7 MW of electricity. In order to cater this demand, the designing of a new capacity recovery boiler is must. As the capacity of the furnace increases to ~4000–5000 TPD, other modifications need to be incorporated such as superheater tubes, economizer and air level system. Increasing the dry solid content in the furnace gives an extra load on the superheater and therefore there is high chances to increase the plugging and corrosion in the tubes. The corrosion and plugging in the tube reduces the heat transfer coefficient and will affect the efficiency of the process. The non process elements entered with the dry solids such as chloride, potassium will increase the corrosion in the boiler bank and tubes. All the superheaters should be kept behind the bullnose in order to avoid the direct contact with the furnace radiation. As heat transfer coefficient increases, the steam temperature increases and this will lead to generate more electricity with high steam pressure.

6.1 Current Trends and Conclusions

Several modifications have been implemented in the recovery furnace in order to cater the need of the industry. The boiler can be modified from two drum to one drum system and provide the water screen. High temperature and pressure steam generation is important these days and therefore, it is required to design the furnace in such a way that it can generate ~9–10 MPa and 480–500 °C. Higher dry solids concentration upto 80% should be fired in the recovery boiler. Liquor should be preheated before being into the furnace by using flash tank. The black liquor concentration should be further increased to 90% by elevating the steam pressure. Alternative materials (biomass) need to be fired such as biological effluent treatment sludge and other waste biomass. In order to control the odor and emission of SOx and NOx, the air level system need to be modified.

Therefore, it is clear that the chemical recovery furnace provides numerous benefits for the effective combustion of black liquor in pulp and paper mills. It facilitates efficient energy recovery by transforming the organic constituents of black liquor into heat and vapor for power production. The furnace facilitates chemical recovery by reclaiming inorganic culinary chemicals for reuse, thereby decreasing raw material costs. It enhances the overall process efficiency, reduces fuel consumption, and refuse discharge. By combining energy generation with chemical recycling, the recovery furnace supports sustainable, cost-efficient, and environmentally conscious mill operations.

References

1. Adams TN, Frederick WJ (1988) Kraft recovery boiler physical and chemical processes. American Paper Institute, Inc., New York, NY (USA)
2. Vakkilainen EK EKBT-SG from B (ed) (2017) Recovery boiler. Butterworth-Heinemann, pp 237–259
3. Blasiak W, Tao L, Vaclavinek J, Lidegran P (1997) Modeling of kraft recovery boilers. Energy Convers Manag 38:995–1005
4. Fujisaki A, Takatsuka H, Yamamura M (1994) World's largest high pressure and temperature recovery boiler: automatic control in a key feature. Pulp Pap Can 95:46–53
5. Donda DP, Simatupang DF (2023) Determination of sodium sulfate requirement for obtaining sodium sulfide content in green liquor at the recovery boiler unit in PT XYZ North Sumatra
6. Pola L, Collado S, Oulego P, Díaz M (2022) Kraft black liquor as a renewable source of value-added chemicals. Chem Eng J 448:137728
7. Gullichsen, J., & Paulapuro, H. (Eds.). (1998–2000). Papermaking Science and Technology (Vols. 1–19). Helsinki, Finland: Fapet Oy.
8. Niemela K (2004) Early history of black liquor research: recovery of by-products andcooking chemicals. Int Recover Boil Conf Porvoo 7–14
9. Sliger RN, Berz MD, Kramlich JC, Malte PC (1996) NOx control in black liquor fuel gas combined cycles. American Society of Mechanical Engineers
10. Wu SL, Lisa K (1998) Kinetics of NO Reduction by Black Liquor Char. Energy & Fuels, 12(3):457–463. DOI: 10.1021/ef970086q
11. Vakkilainen E (2014) Recovery boilers-history and future. Contin Dev Recover Boil Technol years Coop Finl
12. Vakkilainen E (2005) Kraft recovery boilers–principles and practice
13. Mullinger P, Jenkins B (2022) Industrial and process furnaces: principles, design and operation. Butterworth-Heinemann
14. In the context of industrial and environmental engineering as of 2026, JHM primarily refers to the technologies and systems provided by Qingdao Jinhuimo International Trading Co., Ltd. (or Jinhuimo), which specializes in water recovery and treatment plants, and Joule-Heated Melters (JHM) used in waste recovery and vitrification
15. Cardoso M, de Oliveira ÉD, Passos ML (2009) Chemical composition and physical properties of black liquors and their effects on liquor recovery operation in Brazilian pulp mills. Fuel 88:756–763
16. Vakkilainen EK, Ahtila P (2011) Modern method to determine recovery boiler efficiency. O Papel, 72(12):58–65
17. Järvinen M, Zevenhoven R, Vakkilainen E, Forssén M (2003) Black liquor devolatilization and swelling-a detailed droplet model and experimental validation. Biomass Bioenergy 24:495–509
18. Roberts W, Lu H, Ip LT, Baxter LL (2009) Black liquor particle and droplet reactions: experimental and model results. In: 45th anniversary international recovery boiler conference, Lahti, Finland. pp 57–100
19. Ortiz IM, Domènech X, Rieradevall J (2003) Life cycle assessment as a tool for green chemistry: application to Kraft pulp industrial wastewater treatment by different advanced oxidation processes

20. Järvinen MP (2002) Numerical modeling of the drying, devolatilization and char conversion processes of black liquor droplets. Helsinki University of Technology
21. Alén R (1994) Swelling behaviour of kraft black liquor and its organic constituents. Bioresour Technol 49:99–103
22. Chen C (2017) Combustion behavior of black liquors: droplet swelling and influence of liquor composition. Research report/Department of Chemistry, University of Jyväskylä
23. Chen C, Alén R, Lehto J, Pakkanen H (2016) Combustion properties of birch (*Betula pendula*) black liquors from sulfur-free pulping. J Wood Chem Technol 36:401–411
24. Tavares AJ, Tran H, Reid TP (1998) Effect of char bed temperature and temperature distribution on fume generation in a kraft recovery boiler
25. Sutinen J, Karvinen R, Frederick WJ (2002) A chemical reaction engineering and transport model of kraft char bed burning. Ind Eng Chem Res 41:1477–1483
26. Sricharoenchaikul V, Hicks AL, Frederick WJ (2001) Carbon and char residue yields from rapid pyrolysis of kraft black liquor. Bioresour Technol 77:131–138
27. Shires, G. L. (2011) Furnaces. In: THERMOPEDIA™: A-to-Z Guide to Thermodynamics, Heat & Mass Transfer, and Fluids Engineering. Danbury, CT: Begell House, Inc. DOI: https://doi.org//10.1615/AtoZ.f.furnaces
28. The article by R.J. Sarjant: Sarjant RJ (1950) furnance design and practice. Proc Inst Mech Eng 162:75–92
29. Mäkipää M, Backman R (1998) Corrosion of floor tubes in reduced kraft smelts: Studies on effects of chlorine and potassium. In: 9th International symposium on corrosion in the pulp and paper industry
30. Blough JL, Seitz WW, Girshik A (1998) Fireside corrosion testing of candidate superheater tube alloys, coatings, and claddings-phase 2 field testing. Oak Ridge National Laboratory (ORNL), Oak Ridge, TN (United States); Foster Wheeler

Production of Bioethanol as a By-Product in the Pulp and Paper Industry

Ajay Kumar Chauhan, Shailja Chauhan, Aman Singh Rajpoot, and Amit Kumar Chaurasia

Abstract As the world strides for renewable energy solutions, the integration of pulp and paper industries into the bioethanol production landscape emerges as a promising avenue. The pursuit of sustainable and eco-friendly alternatives have led to a paradigm shift in the biofuel landscape, with focus on leveraging the vast potential of pulp and paper industries for bioethanol production. The chapter delves into the innovative approach of utilizing lignocellulosic biomass (LCB), the primary raw material in paper production, as feedstock for bioethanol generation. By exploring the conversion of LCB into simpler sugars, researchers aim to establish a dual-purpose system, yielding both high-value paper products and bioethanol. This chapter provides a comprehensive overview of the ongoing efforts, shedding light on the key developments and challenges face towards making lignocellulosic bioethanol into commercial viable reality. Emphasizing critical role of pretreatment process hydrolysis in the production process, the chapter underscore the significance of this enzymatic or chemical breakdown of LCB as a primary key step for biofuel generation. The exploration of various hydrolysis techniques and their implications on bioethanol yield forms, a central theme, offering insights into optimizing this pivotal stage for enhanced efficiency and economic viability.

Keywords Bioethanol · Lignocellulosic biomass · Hydrolysis · Pretreatment · Pulp and paper

A. K. Chauhan (✉)
Sharda School of Bioscience and Technology, Department of Biotechnology, Sharda University, Greater Noida 201310, Uttar Pradesh, India
e-mail: ajay.chouhan@sharda.ac.in

S. Chauhan
The Rani Laxmi Bai Central Agricultural University, Jhanshi 284003, Uttar Pradesh, India

A. S. Rajpoot
Mechanical Engineering Department, PDPM Indian Institute of Information Technology, Design and Manufacturing, Jabalpur 482001, Madhya Pradesh, India

A. K. Chaurasia
Amity Institute of Biotechnology, Amity University Uttar Pradesh, Noida 201313, Uttar Pradesh, India

S. Sundaramurthy et al. (eds.), *Black Liquor Waste Management in the Pulp and Paper Industry*, Sustainable Materials and Technology,
https://doi.org/10.1007/978-981-96-4237-3_7

Abbreviation

Lignocellulosic biomass	LCB
1G	First generation
2G	Second generation
C5	Pentosan sugar
C6	Hexose sugar
BL	Black liquor
HMF	5-Hydroxymethyl furfural
NaOH	Caustic
ASA	Accessible surface area
SSA	Specific surface area
SScF	Simultaneous saccharification
SHF	Separate hydrolysis and fermentation
SmScF	Simultaneous saccharification and co-fermentation
DMC	Direct microbial conversion
CBP	Consolidated bioprocessing

1 Introduction

The extensive worldwide use of conventional fossil fuels releases greenhouse gases into the environment, and their depletion is the serious concern. To address this problem, researchers are increasingly interested on bio renewables, such as sugar, starches, and lignocellulosic biomass (LCB). LCB are abundantly present on the Earth, and due to its high cellulose and hemicellulose content, it can be converted into simpler sugars, which can serve as the ultimate raw material for biorefinaries. The current consumption of conventional energy has led to debates over food versus fuel debate for 1G ethanol production, prompting researchers to seek effective renewable energy raw materials like LCB. The pulp and paper industry has always been tightly coupled with bioenergy due to the use of wood as feedstock, even as paper and board recycling rates increase. However, the production of paper-based products is expected to continue growing to meet the current demand. Pulp and paper industries can play a significant role by converting their surplus product into a biorefinery conversion process. The development of bio based refineries mainly, second generation (2G) based on conversion of LCB is preferred over first generation (1G), which relies on food staple uses [4]. The production of ethanol from wooden pulp from paper industry can be an alternate option for bioenergy production. The ethanol can be a byproduct of paper industry, which can significantly contribute for increasing a country's economy.

Lignocellulosic biomass (LCB) are found in the form of woods (hardwood, and softwood), grasses, and agricultural wastes (such as wheat straw, rice straw, and paddy stubble). The processing of LCB yields intermediate that can be used for bio-ethanol production. Researchers worldwide are primarily focusing on ethanol production from LCB, and the processes are still in the developmental stage to make them cost-competitive. Nevertheless, the goal is to fulfil global demand for ethanol blending with petroleum, contributing to clean energy and prompting a shift towards a sustainable bio-based economy within the circular economy framework.

Ethanol, as a fuel, provides equivalent combustion efficiency to petrol at lower production cost compared to import, with the key advantage of reducing carbon emissions. Currently, this demand is met through surplus sugarcane and food grain production [4]. It's projected that global bioethanol production in 2022 will reach an expected 180 billions litres [31]. Notably, the Indian Government has decided to increase ethanol blending with fuel from 10 to 20% (*v/v*) by 2025 (NITI aayog, India). Around the world, countries that export petroleum are prioritizing ethanol-blended petrol to meet their energy demands efficiently. LCB rich in cellulosic and hemicellulosic content, consisting C6 and C5 sugars, is the actual composition of pulp and paper. Due to their structural composition of pulp and paper, ethanol can be produced using an enzymatic or microbial fermentation, offering a valuable alternate to gasoline [34]. LCB-based ethanol production generally involve pretreatment as the first step to break down recalcitrant lignin structure. The leftovers after LCB pretreatment, mainly contains lignin, which can be further processed into value added product [7]. Further, improvements in terms of yield from cellulose and hemicellulose, fractionation of C5 and C6 sugars or co-catabolism and simultaneous sugar utilization can be applied to increase the saccharification, which is an integral part of biochemical processing of LCB [11]. Further, suitable fermentation process converts these sugars into ethanol.

To meet the growing demand for clean energy, pulp and paper industry can make a significant contribution. Paper primarily consists of cellulose, hemicellulose and traces of lignin depending on the type of pulp and bleaching process. The production of cellulose and hemicellulose, key components of paper, is primarily produced by the pulp and paper industry. The surplus production of these cellulose and hemicellulose materials hold great potential for the pulp and paper based bio-refineries. These polymers are rich sources of hexose (C6) and pentose (C5) sugars. Therefore, in this book chapter, the production cellulosic and hemicellulosic ethanol is discussed in detail, covering the transition from pilot-level to commercial-scale production techniques for processing LCB for bioethanol production.

1.1 Chemical Composition of Lignocellulose Biomass

Considering the current energy demand, LCB has gained global attention as a sustainable and environmentally friendly source of bioenergy. Due to it's chemical composition, LCB can act as an alternative to fossil fuels and play a pivotal role in second generation (2G) biofuels generation. Woody biomass and associated waste materials, including agricultural residues (stubbles, paddy straw, wheat straw), forest wastes (pine waste, plant based biomass), miscantus, switch grass, poplar, and others, which are the main raw materials for paper manufacturing, can also be effectively utilized for biorefinery applications [46].

In LCB, cellulose constitutes a significant portion, ranging from 40 to 60% by weight, with cellobiose units predominantly linked by β-(1–4). This sugar polymer is present in crystalline and amorphous form [8]. Hemicellulose, on the other hand, accounts for approximately 25–30% of the LCB's weight and consists of various heterogeneous biopolymers, primarily containing C5 sugars such as xylans, xyloglucan, and mannans and other sugars. Hemicellulose has considerably lower degree of polymerization compared to cellulose. In pulp and paper based refineries, hemicellulose acts as a physical barrier after lignin, necessitating pretreatment methods like dilute acid or base treatment, steam explosion, auto hydrolysis, and others to increase the accessibility of cellulosic content for enzymatic processes.

Furthermore, the acetylation of hemicellulose in the LCB significantly hinders the accessibility of cellulosic accessibility for its conversion into sugar using chemical or biochemical methods [37]. After cellulose and hemicellulose, lignin constituents the highest reported content in LCB, ranging from 10 to 32% in it. Pulp and paper industries have well-established processes such as Kraft, soda, and acid pulping to separate cellulosic fibers, removing lignin and converting it into paper [7]. Lignin in LCB possesses recalcitrant structure with β-O-4 linkages, makes it challenging for effective removal from LCB, therefore, suitable combination of chemical and biochemical pretreatment method can solve the problem. Alternatively thermochemical processes, including gasification, pyrolysis, and hydrothermal carbonization, are employed to produce valuable chemicals from lignin [13].

Presently, lignin in the form of black liquor (BL) is utilized by pulp and paper industries for co-power generation. However, this lignin in BL can also be harnessed for the production of aromatics and linear chain hydrocarbons, reducing the industry's dependence on petrochemical sources for fuel requirement [13]. Due to the recalcitrant nature of carbon compounds present in LCB, processing becomes more challenging and intriguing. Consequently, suitable pretreatment processes are necessary to achieve cost-competitive conversion of consolidated LCB processing into desirable product.

1.2 Pretreatment of LCB

Pretreatment stands as a crucial unit operation in pulp and paper based biorefineries. An overview of the detailed process of 2G ethanol production from LCB is presented in Fig. 1. The nature and chemical composition of feedstocks are essential parameters for determining the pretreatment processes. The primary purpose of LCB purpose of pretreatment is to reduce cellulose crystallinity, thereby increasing the exposed surface area in comparison to untreated biomass. This process involves the removal of lignin and the enhanced recovery of cellulose and hemicellulose for further conversion.

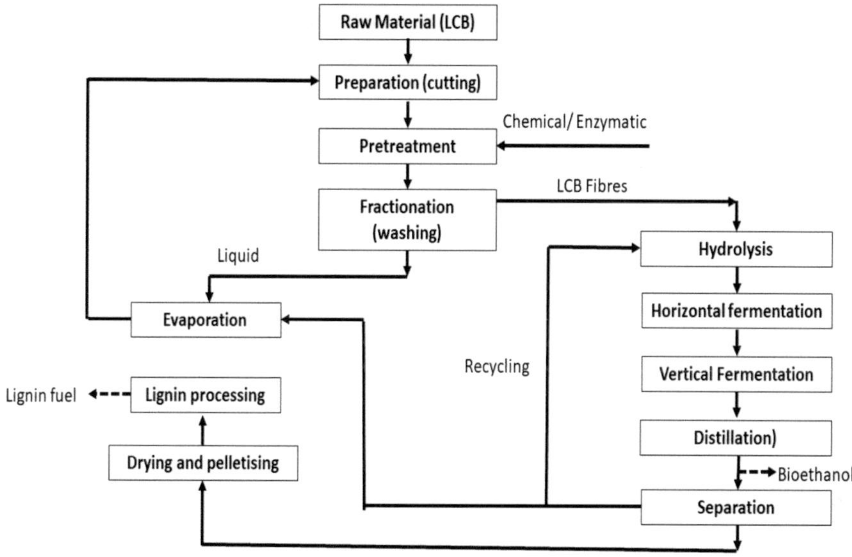

Fig. 1 Process overview of lignocellulosic biomass based biorefineries

LCB, cellulose, which accounts for 30–50% of its composition, primarily consists of homopolysaccharide containing β-D-glyosidic bonds (7000–15,000 carbon chain). Hemicellulose, on the other hand, (100–3000 carbon chain) has shorter chain length and is a branched polysaccharide. Hemicellulose can be more easily hydrolyzed using dilute acid and low concentration of enzymes [40]. The removal of lignin significantly reduces the overall recalcitrant nature of LCB. In the current time, there are different types of pretreatment methods available, categorized as physical, chemical, and biological depending upon their treatment approach. Additionally, the fractionation of sugar molecules (C6 and C5) and lignin monomers presence is a challenging task. The conversion of C6 sugars into hydroxymethyl furfural (HMF), and C5 into furfural, and lignin into lignin oils or fillers pose significant challenges in the further processing of LCB [10]. These aspects are discussed in detail in the following section. Previous studies have shown that deacetylation of various LCB types, such as wheat straw, poplar, grasses and switchgrass contributes to improved hemicellulose digestibility compared to cellulose [46].

1.3 Physical Treatment

1.3.1 Hydrothermal Pretreatment

The physical treatment methods majorly involve the use of physiological parameters such as elevated temperature and pressure to remove lignin from LCB. In hydrothermal pretreatment, as the name suggests, involves that addition of water for the pretreatment of LCB and it is also known as the autohydrolysis process. These processes are green, clean and sustainable, causing no harm to the instrument in terms of corrosion. However, to improve their efficiency, it can be coupled with chemical/ or biological methods. In combinatorial methods with acidic pretreatment, addition of sulphuric acids is generally required, which can cause damage to the equipment due to corrosion and lead to sugar oxidations [3]. In acidic condition, the formation of undesirable by-products such as high level of furfural, 5-hydroxymethyl furfural (HMF), levulenic acid, and formic acid are found in the exit streams [3].

The fractionation process of cellulose and hemicellulose from treated LCB requires a significant capital investment for its fractionation processing. Therefore, an alternate cost effective alkaline pretreatment process is suggested for further processing. In case of alkaline treatment (Sulfite-NaOH) of *Eucalyptus* after hydrothermal pre-treatment, followed by enzymatic treatment, it has the potential to achieve a glucose conversion efficiency of 44–78%. The leftover from sulfite-NaOH results in a 21% yield of sulfonated lignin [3]. The treated sugar part (ethanol production) and obtained lignosulfonates (energy generation) could serve as a source of bio renewables.

1.3.2 Autohydrolyis

The concept of auto hydrolysis of hardwood is an efficient way for converting woody biomass into ethanol producing intermediates. In this context, a pretreatment step is necessary to increase enzymatic accessibility for converting cellulose and hemicellulose into simpler sugars. Earlier, autohydrolysis processes were not recommended due to the low yield of product formation. However, the limitation of autohydrolysis is overcome by modifying refining as the first initial step, and increasing the available surface area for enzymatic action. Consequently, the efficiency was significantly improved (25%) by refining [27]. The other way is to control the formation of undesirable byproducts such as furfural and acetic acid. This can be achieved by controlling prehydrolysis in the fiber and water system. Increasing pulp consistency using a screw press in the operation helps in removal of water and makes the process cost-competitive. The screw press aids in removing large quantities of dissolved solids containing furfural and acetic acid. The separation of undesirable components like acetic acid and furfural could be accomplished using membrane or ion exchange separation. For increasing efficacy, steam stripping of furfural and 5-hydromethyl furfural can help to achieve an autohydrolysis efficiency of 70% [24].

Further, improvement in hydrolysis efficiency could be achieved by modifying the existing process. To make autohydrolysis more efficient, a two-step approach has been suggested [10]. This process uses surplus green liquor and makes process economical, similar to modifying conventional caustization to enhance the Kraft pulping efficiency [32].

1.4 Chemical Pretreatment of Lignocellulosic Biomass

Chemical treatment involves various types of chemicals such as acidic, basic, and oxidative methods, which can be combined with physical methods like steam explosion or hydrothermal treatment, or a combination of these techniques. Additionally, organic solvents, and ionic liquids also helps in removing recalcitrant lignin from lignocellulosic biomass [8]. It is observed that, post lignin removal, fractionation of cellulose and hemicellulose from LCB is a cost intensive process. The fractionation methods currently used in LCB pretreatment can significantly alter lignin's structure. However, preserving β-O-4 linkages in lignin can help in conducive fractionation of lignin. Studies on pretreatment have shown suitable selection of process makes it more cost competitive. For instance, alkaline pretreatments combined with sodium sulfite (Na_2SO_3) are notable, as it has the capacity to extract sulfonate lignin, making cellulose and xylan more accessible to enzymes [42]. From the structural standpoint of view, lignin and hemicellulose with the cellulosic unit involve both covalent and non-covalent bonds, resulting in a more intriguing structure. Therefore, dilute acid treatment has been suggested and in a previous study, sugar yields up to 90%, including a mixture of C5 and C6 has been obtained [10]. The effect of chemical used for the pretreatment process, hydrolysis and post hydrolysis time, temperature, pH, with effective solid loading optimization are the key for making the process economical viable. The most efficient and cost-competitive approach often involves a combination of chemical, physical, and biological methods of pretreatment. The use of chemical methods alone can release of some toxic environmental pollutants, and its removal adds to the overall cost of pretreatment. Therefore, there is a growing need for green and sustainable technologies that promote a circular economy while protecting the environment.

1.4.1 Acid Pretreatment

Acid pretreatment methods involve the use of organic and inorganic acid for the pretreatment of LCB. From the previous literature, sulphuric acid is preferably used for the pretreatment compared to other acid pretreatment processes [14]. In an optimized condition of 15% solid loading (*w/w*) of bamboo, 5% acid (*v/v*) for 30 minutes yield 0.319 g/g of reducing sugar [2]. The major limitation of using sulphuric acid for pretreatment is the generation of furfural as an intermediate, which further hinders the fermentation process. Therefore, changes in the use of alternate method like different acids could be the better alternative. For instance, Koostra et al. studied organic and inorganic acids uses on the pretreatment process of LCB, and concluded that the use of organic acid could be a better alternate [21]. They also suggested that organic acids are less corrosive and cost competitive compared to sulphuric acid, and does not produce furfural. For the better yield of sugar in acid hydrolysis, generally two steps are involved: the first step is deconstruction of LCB, and second is post hydrolysis conversion of amorphous cellulose in glucose. The use of dilute acid is advantageous in terms of no recovery of used acid, but it has limitation of high temperature (180 °C) and pressure (1 atm) [20].

1.4.2 Alkaline Pretreatment

Alkaline pretreatment is performed at pH 9.0-12.0 conditions. The various alkaline chemicals have been used for alkaline pretreatments, which are mostly hydroxides of calcium, potassium, sodium and ammonium. Among all the methods, NaOH treatment is found to be the most effective [17]. Saponification reaction helps in breaking of hemicellulose and the linkage with lignin. The key advantage of alkaline pretreatment yields higher fermentable sugars compared to other pretreatment processes (Table 1).

1.5 Biological Treatment

The key limitation of physiochemical pretreatment methods are the highly energy intensive and toxic byproducts formation, could be responsible for low saccharifcation efficiency. These limitations are the opportunity for biological process researchers to develop cost competitive, energy efficiency and sustainable process. Biological pretreatment process involve the use of microbial communities and their enzymes for the degradation of lignin. These enzymes are primarily lignocellulolytic enzymes of fungal origin, often obtained from white and brown rot fungi [9]. These microbial species and their enzymes play a dual role in degrading lignin while simultaneously hydrolyzing the cellulose and hemicellulose content of biomass [28]. Biological processes offer distinct advantages over chemical and physical treatment processes due to their lower energy consumption and reduced formation of toxic byproduct and inhibitors. The limitation of physiochemical processes have spurred researchers to explore biological alternatives more widely.

1.5.1 Microbial Treatment

Various microorganisms have the potential to treat the cellulosic biomass, which is the initial step in the bioethanol production. Some of these microorganism are listed in Table 2 and are valuable for the microbial degradation of LCB. Microbial treatment methods are environmentally sustainable, but one major limitation is a fact that the degradation processes can be time consuming. In the microbial degradation process, lignin is primarily broken down by the different classes of fungi. These fungi, include ascomycetes, basidiomycetes, and deuteromycetes. One well-known group of fungi that plays a significant role in lignin degradation is the "White rot fungus". Additionally, some bacterial species have also been used, which secretes extracellular enzymes (mainly cellulases and lignolytic enzymes) [5]. These enzymes

Table 1 Various chemical treatment processes

Catalysts	Reactions types and conditions	Effects	Remarks	References
Sodium hydroxide-ultra sound	0.25–2.5 M NaOH, 50–90 °C, power (40–140 W), frequency 20 kHz, 1 h, 1:20 solid loading	Delignification of 67.3% of sugarcane bagasse	High reaction rate, cost competitive compared to conventional ultrasonic bath	[18]
Sodium Hydroxide	2% NaOH, 80 °C, Solid loading 10% *w/v*	Herbaceous and woody biomass, Lignin removal 16–39.6%	Helps in increasing sacchrification efficiency of enzyme	[41]
Sodium carbonate	1–30% Na_2CO_3, 60–180 °C, 5–60 min solid loading 10–30%, corn stover, *Miscanthus*, and swithcgrass	40–59% delignification, increased glucose and xylose yield	Preserve most of the cellulosic content, increased saccharification efficiency of enzyme	[25]
Two step hydrothermal treatment and Aqueous ammonia extraction	Hydrothermal 5 h, 140 °C, 12% aqueous ammonia extraction 1–2 h at 100–160 °C, wheat straw	10–50% hemicellulase solubilisation, 0–80% delignification Profound swelling, lignin removal or modification	Pentosan content decreased and extraction of lignin increased to 15% and total carbohydrate of 78.1%	[33]
Lime	0.1 g/g, 120 °C, 1 h, 1.25 kPa, 0.05–0.15 g $Ca(OH)_2$ solid loading 5–20%	30% lignin, 5% hemicellulose, 11% cellulose accumulation, and 70% digestion of the cellulose content	Simple and low cost process, gives cellulose rich component	[15]

are capable of lignin degradation and the simultaneous degradation of cellulose and hemicellulose.

1.5.2 Enzymatic Treatment

An enzymatic intervention into the pulp and paper industries is an environmental friendly, sustainable and carbon neutral approach. Due to environmental pressure, biological treatment methods are preferred over thermo-chemical and chemical treatment methods. In this context, cellulolytic and hemicellulytic enzymes produced by *Trichoderma reesei* are widely studied for biomass bioconversion. However, cheaper substrate feedstocks are required for the production of these enzymes, which

Table 2 Various microbial strains used for LCB degradation

Microorganism	Biomass	Effects	References
Punctualaria sp. TUFC20056	Bamboo culms	Up to 50% lignin removal	[35]
Irpex lactueus	Corn stalks	Up to 82% hydrolysis yield	[12]
Fungal consortium	Straw	Seven times increase in hydrolysis	[36]
P. ostreatus/P. pulmonarius	*Eucalyptus grandis* saw dust	Twenty times increase in hydrolysis	[9]
P. chrysosporium	Rice husk	447.75 mg/mL.g	[28]
Pyrenophora phaeocomes	Rice straw	470 g/g, 50–51% of lignin and hemicellulose removal	[29]

can produce new route of biorefinery for fulfilling the energy demand. As we know, cellulase converts cellulose into the simpler C6 sugar. If the production and separation cost of the enzyme or stable crude enzyme can reduce the operational cost which can ultimately help in lowering the energy demand. Presently, the ethanol production via LCB is not an economical option because several intermediate steps are required to convert LCB into sugar. However, for making an enzymatic step more efficient for hydrolysis of LCB, extensive care should be taken to make the enzyme active throughout the catalysis process.

2 Factors Influencing LCB Hydrolysis

The physiochemical parameters, which significantly influences the hydrolysis of LCB, are pH, temperature, and pressure. The most efficient experimental conditions itself depends upon the raw material used for pulp production.

2.1 Cellulose

It has been observed that cellulose fibrils are embedded in the LCB matrix, making enzymatic hydrolysis stringent. Conversely, the presence of cellulose in LCB exhibits a high degree of polymerization (DP) in the range of 500–1400 D-glucose, which can be directly related to crystallinity [46]. Higher crystallinity imparts recalcitrance to the cellulose structure due to strong non-covalent hydrogen bonding. Therefore, the primary challenge is to overcome recalcitrance induced by cellulose. On the contrary, amorphous structure of cellulose facilitates the availability of the surface area for the better conversion of cellulosic unit into simpler sugar [1]. The degree of polymerization and sugar crystallinity are the important parameter that determines the reaction rate of hydrolysis. The cellulase enzyme possesses the potential to deconstruct crystalline and amorphous cellulose, which is the key benefit of this enzyme.

2.2 *Hemicellulose and Lignin*

The presence of hemicellulose and lignin in LCB is responsible for the low conversion of hydrolysis. Hemicellulose, with a 100–200 DP and an amorphous in nature, has lower strength compared to lignin and cellulose, making it to be easily hydrolyzed by dilute acid/base/hemicellulases treatment [46]. Hemicellulose act as a physical barrier that can limit the accessibility of enzymes. This limitation can be overcome by dilute acid and steam explosion treatment [30]. On the other hand, lignin is a highly amorphous hetero-polymer and composed of highly complex phenylpropanoid unit [8]. Lignin provides structural rigidity to the LCB and serves as a physical barrier, blocking the accessibility of cellulose to the enzymes and chemicals. Therefore, the removal of hemicelluloses and lignin can increase the pore size and provide higher surface area for enzymes to act, converting cellulosic unit into simpler sugar [46].

2.3 *Accessible Specific Surface Area and Accessible Volume*

Accessible surface area (ASA) is a critical factor that significantly improves the hydrolyzing efficiency. It is directly related to porosity, specific surface area (SSA) and pore volume [23]. The reduction in particle size or increase in the pore volume of LCB leads to a higher surface area available for enzyme binding. However, it's important to note that reducing the size is an energy intensive process. Therefore, reasonable size of the raw material is recommended. For practical purpose, determining ASA can be challenging, but SSA is more commonly used to assess the available surface area for accessibility of enzymes. Moreover, accessible volume, controlled by pore size, is a crucial factor influencing enzymatic conversion of cellulose into sugars. Therefore, accessible volume of cellulose in LCB plays a pivotal role in achieving higher enzymatic conversion [16].

2.4 *Prediction of Enzymatic Hydrolysis*

Enzymatic catalysis is a multivariant and multiscale phenomenon controlled by various physiochemical process parameters, including lignin, hemicellulose and cellulose and their respective degree of polymerization, as well as accessible surface area. The presence of various substances such as cellobiose, glucose, xylose, HMF, non-desirable intermediates, and substrate surface charges can inhibit the enzyme activity [43]. The addition of β-glucosidase has been shown to synergistically reduce the inhibition of cellobiose [45]. However, higher content of glucose inhibit not only the β-glucosidase but also cellulase [19]. To enhance the hydrolysis efficiency, a potential strategy involves the removal of hydrolyzed sugars during the processes.

2.5 Mixing

Suitable mixing during hydrolysis is crucial for reducing heterogeneity and increasing the conversion of pulp from LCB into fermentable sugar [22]. The design of reactor and its agitator plays a significant role in effectively handling of mixture and mixing inside the reactor [6]. Zhang et al. [44] designed a reactor equipped with double helical impeller featuring an S-shaped design [44]. The use of this impeller resulted in the best yield due to improved mixing, and it also reduced the overall process energy demand. Moreover, they reported that, at 30% solids loading, the system could operate smoothyly. This type of impellar is employed for fiber homogenization to enhance mixing. A Plus bioreactor (B. Braun Biotech International, Germany), equipped with a pitched blade impeller with three blades at an angle of 45° (diameter of 70 mm and a blade width of 20 mm), were used for the hydrolysis. At different speeds, up to 600 rpm, the hydrolysis increased by 37% [22].

3 Microbial Conversion of Sugar into Ethanol

The user friendly host such as *Zymomonas mobilis*, *E. coli*, and *Clostridum* have been the focal point of recent metabolic engineering studies aimed at developing a model microbial chassis for sustainable biorefinery practices in the modern era. Microbes, including yeast, *E. coli*, *Zymomonas*, capable of converting both C5 and C6 sugars into ethanol, face significant limitations. Therefore, there is a pressing need to create a consortium of microbial populations that can simultaneously convert both sugars into ethanol. The use of synthetic biology, involving genetic manuipulation of genes, can facilitate the co-utilization of sugars and enhance the yield of ethanol from the carbon sources present in the medium. By employing synthetic biology tools, microbial strains can be developed to efficiently consume sugars and convert them into ethanol.

$$C_6H_{12}O_6 \rightarrow 2C_2H_5OH + 2CO_2$$

4 Saccharification

4.1 Simultaneous Saccharification and Fermentation

In SScF, enzymatic hydrolysis and fermentation of glucose and xylose occur in the same reactor. Enzymatic hydrolysis, either through externally added enzymes or those produced by the microorganism, begins after pretreatment, allowing for completion in a single vessel with high productivity. However, the process is limited to hexose sugars, there's minimal provision for pentose sugars. Despite these limitations, SScF offers advantages such as reduced capital input, continuous removal of end products to reduce enzyme inhibition, and operation at higher concentrations of insoluble solids for increased product concentration.

4.2 Separate Hydrolysis and Fermentation (SHF)

It is a conventional fermentation method for LCB. In SHF, biomass is treated separately in a reactor with mineral acids or base to hydrolyze sugars. These sugars are then fermented separately in a reactor using conventional methods optimized with batch, fed batch and continuous fermentation. While SHF has the advantage of separate vessel operation, it faces limitations such as enzyme inhibition during sugar hydrolysis and high input costs and time consumption in the fermentation process. Moreover, this process of fermentation requires extensive cost and prolonged fermentation time [38].

4.3 Simultaneous Saccharification and Co-Fermentation (SmScF)

SmScF focuses on hydrolyzing both hexose and pentose sugars simultaneously or sequentially [39]. Enzymes breakdown lignocellulosic biomass into hexose and pentose sugars and glucose to xylose ratio are maintained for optimal fermentation. Co-fermentation in a single reactor allows for simultaneous or sequential utilization of both sugars. This process offers the advantage of collecting products from both sugars after fermentation, enabling direct downstream processing. Challenges include maintaining proper glucose to xylose ratios to prevent substrate inhibition and monitoring for secondary metabolites.

5 Direct Microbial Conversion (DMC)/Consolidated Bioprocessing (CBP)

DMC/CBP involves simultaneous enzyme production, enzymatic hydrolysis, and fermentation) [26]. This approach reduces the production costs by employing cellulytic and fermenting microbes. Research examples include a metabolic engineered strain of *C. cellulvorans* DSM743B achieving significantly higher butanol production compared to the wild type strain-using corncob as a substrate. Genetically modified strains with enhanced cellulose production show improved glucose concentration and ethanol yield from corncob cellulose in fermentation, demonstrating the potential of DMC/CBP in lignocellulose biomass conversion.

6 Conclusions and Future Perspective

In the quest for a sustainable future, the pulp and paper industry must adopt holistic approach that integrates technological innovations, and environmentally conscious practices. A commitment to extracting maximum value from every component of lignocellulose biomass is paramount. Strategic advancements and a dedication to continuous and resource-efficient global economy are pivotal. One key avenue lies in enzymatic hydrolysis, which holds tremendous potential for the industry. Future prospects unfold in several areas, shaping the trajectory of the pulp and paper sector.

1. **Efficient Raw Material Utilization**

The integration of advanced enzymatic hydrolysis techniques can enhance the efficient utilization of raw materials in the pulp and paper industry. Tailoring enzymes to the specific characteristics of cellulose, hemicellulose, and lignin can lead to improved yields and cost-effective processes.

2. **Reduced Environmental Impact**

Exploration of green pretreatment methods holds promise for reducing the environmental impact of pulp and paper production. Sustainable alternatives, such as biological pretreatments or environmentally friendly solvents, align with the industry's growing focus on eco-friendly practices.

3. **Integrated Biorefineries in Paper Mills**

The concept of integrated biorefineries within paper mills can revolutionize the industry. By combining enzymatic hydrolysis with traditional pulping process, paper mills can extract additional value from LCB, producing not only paper but also biofuels and biochemical.

4. **Energy Efficiency Measures**

Considering energy, the energy intensive nature of size reduction process, finding energy-efficient alternatives and technologies becomes imperative.

Understanding and optimizing the factors influencing enzymatic hydrolysis of lignocellulosic biomass (LCB) is crucial for advancing bioconversion processes. An efficient enzymatic hydrolysis of LCB remains a critical challenge in the development of sustainable biofuel and biochemical production. Strategies to enhance the surface area, including reducing particle size and increasing pore volume, offer avenues for improvement. However, balancing these strategies with energy efficiency considerations is essential. As we continue to unravel the complexities of LCB and enzymatic hydrolysis, interdisciplinary collaborations and innovative approaches will be pivotal in driving the bioconversion field towards achieving more sustainable and economically viable solutions.

References

1. Alvira P, Tomás-Pejó E, Ballesteros M, Negro MJ (2010) Pretreatment technologies for an efficient bioethanol production process based on enzymatic hydrolysis: a review. Bioresour Technol 101(13):4851–4861. https://doi.org/10.1016/j.biortech.2009.11.093
2. Baruah J, Nath BK, Sharma R, Kumar S, Deka RC, Baruah DC, Kalita E (2018) Recent trends in the pretreatment of lignocellulosic biomass for value-added products. Front Energy Res 6:141. https://doi.org/10.3389/fenrg.2018.00141
3. Cedeno FRP, Siqueira BB, Chavez EGS, Roldán IUM, Ropelato LM, Galán JPM, Masarin F (2022) Recovery of cellulose and lignin from Eucalyptus by-product and assessment of cellulose enzymatic hydrolysis. Renew Energy 193:807–820. https://doi.org/10.1016/j.renene.2022.05.027
4. Chauhan AK (2020) Biofuel: types and process overview. In: Srivastava M, Srivastava N, Mishra PK, Gupta VK (eds) Nanomaterials in biofuels research. Springer Singapore, Singapore, pp 1–28
5. Chauhan AK, Choudhury B (2022) Lignolytic enzymes and their role in the bioremediation of environmental pollutants: prospects and challenges. In: Microbial remediation of azo dyes with prokaryotes, pp 117–128
6. Chauhan AK, Kalyan G (2021) Types of bioreactors for biofuel generation. In: Srivastava N, Srivastava M, Mishra PK, Gupta VK (eds) Bioprocessing for biofuel production: strategies to improve process parameters. Springer Singapore, Singapore, pp 57–79
7. Chauhan AK, Singh SP (2023) Lignin valorization and generation of precipitated lignin: a sustainable route for circular economy in the context of the Indian pulp and paper industry. Indian Pulp Pap Techn Assoc 35:127–132
8. Chauhan AK, Srivastava SK, Singh S (2021) Lignin depolymerization strategy and role of ionic liquids. In: Srivastava M, Srivastava N, Singh R (eds) Bioenergy research: integrative solution for existing roadblock. Springer Singapore, Singapore, pp 157–173
9. Cianchetta S, Di Maggio B, Burzi PL, Galletti S (2014) Evaluation of selected white-rot fungal isolates for improving the sugar yield from wheat straw. Appl Biochem Biotechnol 173(2):609–623. https://doi.org/10.1007/s12010-014-0869-3
10. Conrad M, Smirnova I (2020) Two-step autohydrolysis pretreatment: towards high selective full fractionation of wheat straw. Chem Ing Tech 92(11):1723–1732. https://doi.org/10.1002/cite.202000056
11. del Río PG, Gullón B, Wu J, Saddler J, Garrote G, Romaní A (2022) Current breakthroughs in the hardwood biorefineries: hydrothermal processing for the co-production of xylooligosaccharides and bioethanol. Bioresour Technol 343:126100. https://doi.org/10.1016/j.biortech.2021.126100

12. Du W, Yu H, Song L, Zhang J, Weng C, Ma F, Zhang X (2011) The promoting effect of byproducts from *Irpex lacteus* on subsequent enzymatic hydrolysis of bio-pretreated cornstalks. Biotechnol Biofuels 4(1):37. https://doi.org/10.1186/1754-6834-4-37
13. Erfani Jazi M, Narayanan G, Aghabozorgi F, Farajidizaji B, Aghaei A, Kamyabi MA, Navarathna CM, Mlsna TE (2019) Structure, chemistry and physicochemistry of lignin for material functionalization. SN Appl Sci 1(9):1094. https://doi.org/10.1007/s42452-019-1126-8
14. Gönen Ç, Deveci EÜ, Akter Önal N (2021) Evaluation of biomass pretreatment to optimize process factors for different organic acids via Box-Behnken RSM method. J Mater Cycles Waste Manag 23(5):2016–2027. https://doi.org/10.1007/s10163-021-01276-7
15. Grimaldi MP, Marques MP, Laluce C, Cilli EM, Sponchiado SRP (2015) Evaluation of lime and hydrothermal pretreatments for efficient enzymatic hydrolysis of raw sugarcane bagasse. Biotechnol Biofuels 8(1):205. https://doi.org/10.1186/s13068-015-0384-y
16. Herbaut M, Zoghlami A, Habrant A, Falourd X, Foucat L, Chabbert B, Paës G (2018) Multimodal analysis of pretreated biomass species highlights generic markers of lignocellulose recalcitrance. Biotechnol Biofuels 11(1):52. https://doi.org/10.1186/s13068-018-1053-8
17. Kim M, Day DF (2011) Composition of sugar cane, energy cane, and sweet sorghum suitable for ethanol production at Louisiana sugar mills. J Ind Microbiol Biotechnol 38(7):803–807. https://doi.org/10.1007/s10295-010-0812-8
18. Kininge MM, Gogate PR (2022) Intensification of alkaline delignification of sugarcane bagasse using ultrasound assisted approach. Ultrason Sonochem 82:105870. https://doi.org/10.1016/j.ultsonch.2021.105870
19. Konar S, Sinha SK, Datta S, Ghorai PK (2019) Probing the effect of glucose on the activity and stability of β-glucosidase: an all-atom molecular dynamics simulation investigation. ACS Omega 4(6):11189–11196. https://doi.org/10.1021/acsomega.9b00509
20. Kong-Win Chang J, Duret X, Berberi V, Zahedi-Niaki H, Lavoie J-M (2018) Two-step thermochemical cellulose hydrolysis with partial neutralization for glucose production. Front Chem 6:117. https://doi.org/10.3389/fchem.2018.00117
21. Kootstra AMJ, Beeftink HH, Scott EL, Sanders JPM (2009) Optimization of the dilute maleic acid pretreatment of wheat straw. Biotechnol Biofuels 2(1):31. https://doi.org/10.1186/1754-6834-2-31
22. Liguori R, Ventorino V, Pepe O, Faraco V (2016) Bioreactors for lignocellulose conversion into fermentable sugars for production of high added value products. Appl Microbiol Biotechnol 100(2):597–611. https://doi.org/10.1007/s00253-015-7125-9
23. Liu Z-H, Qin L, Li B-Z, Yuan Y-J (2015) Physical and chemical characterizations of corn stover from leading pretreatment methods and effects on enzymatic hydrolysis. ACS Sustain Chem Eng 3(1):140–146. https://doi.org/10.1021/sc500637c
24. Marzialetti T, Valenzuela Olarte MB, Sievers C, Hoskins TJC, Agrawal PK, Jones CW (2008) Dilute acid hydrolysis of loblolly pine: a comprehensive approach. Ind Eng Chem Res 47(19):7131–7140. https://doi.org/10.1021/ie800455f
25. Mirmohamadsadeghi S, Chen Z, Wan C (2016) Reducing biomass recalcitrance via mild sodium carbonate pretreatment. Bioresour Technol 209:386–390. https://doi.org/10.1016/j.biortech.2016.02.096
26. Parisutham V, Kim TH, Lee SK (2014) Feasibilities of consolidated bioprocessing microbes: from pretreatment to biofuel production. Bioresour Technol 161:431–440. https://doi.org/10.1016/j.biortech.2014.03.114
27. Phillips RB, Jameel H, Chang HM (2013) Integration of pulp and paper technology with bioethanol production. Biotechnol Biofuels 6(1):13. https://doi.org/10.1186/1754-6834-6-13
28. Potumarthi R, Baadhe RR, Nayak P, Jetty A (2013) Simultaneous pretreatment and saccharification of rice husk by *Phanerochaete chrysosporium* for improved production of reducing sugars. Bioresour Technol 128:113–117. https://doi.org/10.1016/j.biortech.2012.10.030
29. Rastogi S, Soni R, Kaur J, Soni SK (2016) Unravelling the capability of *Pyrenophora phaeocomes* S-1 for the production of ligno-hemicellulolytic enzyme cocktail and simultaneous bio-delignification of rice straw for enhanced enzymatic saccharification. Bioresour Technol 222:458–469. https://doi.org/10.1016/j.biortech.2016.10.012

30. de Oliveira Santos VT, Siqueira G, Milagres AMF, Ferraz A (2018) Role of hemicellulose removal during dilute acid pretreatment on the cellulose accessibility and enzymatic hydrolysis of compositionally diverse sugarcane hybrids. Ind Crops Prod 111:722–730. https://doi.org/10.1016/j.indcrop.2017.11.053
31. Sharma B, Larroche C, Dussap C-G (2020) Comprehensive assessment of 2G bioethanol production. Bioresour Technol 313:123630. https://doi.org/10.1016/j.biortech.2020.123630
32. Singh IJ, Tyagi MK, Ajay C (2012) Modified conventional causticizing (MCC) - a step towards cost reduction 24:111–113
33. Sipponen MH, Österberg M (2019) Aqueous ammonia pre-treatment of wheat straw: process optimization and broad spectrum dye adsorption on nitrogen-containing lignin. Front Chem 7:545. https://doi.org/10.3389/fchem.2019.00545
34. Streva ER, Pasa VMD, Sodré JR (2011) Aging effects on gasoline–ethanol blend properties and composition. Fuel 90(1):215–219. https://doi.org/10.1016/j.fuel.2010.07.056
35. Suhara H, Kodama S, Kamei I, Maekawa N, Meguro S (2012) Screening of selective lignin-degrading basidiomycetes and biological pretreatment for enzymatic hydrolysis of bamboo culms. Int Biodeterior Biodegrad 75:176–180. https://doi.org/10.1016/j.ibiod.2012.05.042
36. Taha M, Shahsavari E, Al-Hothaly K, Mouradov A, Smith AT, Ball AS, Adetutu EM (2015) Enhanced biological straw saccharification through coculturing of lignocellulose-degrading microorganisms. Appl Biochem Biotechnol 175(8):3709–3728. https://doi.org/10.1007/s12010-015-1539-9
37. Tarasov D, Leitch M, Fatehi P (2018) Lignin–carbohydrate complexes: properties, applications, analyses, and methods of extraction: a review. Biotechnol Biofuels 11(1):269. https://doi.org/10.1186/s13068-018-1262-1
38. Toor M, Kumar SS, Malyan SK, Bishnoi NR, Mathimani T, Rajendran K, Pugazhendhi A (2020) An overview on bioethanol production from lignocellulosic feedstocks. Chemosphere 242:125080. https://doi.org/10.1016/j.chemosphere.2019.125080
39. Ullah M, Khattak W, Islam M, Khan S, Park J (2014) Bio-ethanol production through simultaneous saccharification and fermentation using an encapsulated reconstituted cell-free enzyme system. Biochem Eng J 91:110–119. https://doi.org/10.1016/j.bej.2014.08.006
40. Usmani Z, Sharma M, Awasthi AK, Lukk T, Tuohy MG, Gong L, Nguyen-Tri P, Goddard AD, Bill RM, Nayak SC, Gupta VK (2021) Lignocellulosic biorefineries: the current state of challenges and strategies for efficient commercialization. Renew Sustain Energy Rev 148:111258. https://doi.org/10.1016/j.rser.2021.111258
41. Wang W, Wang X, Zhang Y, Yu Q, Tan X, Zhuang X, Yuan Z (2020) Effect of sodium hydroxide pretreatment on physicochemical changes and enzymatic hydrolysis of herbaceous and woody lignocelluloses. Ind Crops Prod 145:112145. https://doi.org/10.1016/j.indcrop.2020.112145
42. Wang W, Zhuang X, Yuan Z, Yu Q, Qi W, Wang Q, Tan X (2012) High consistency enzymatic saccharification of sweet sorghum bagasse pretreated with liquid hot water. Bioresour Technol 108:252–257. https://doi.org/10.1016/j.biortech.2011.12.092
43. Zhai R, Hu J, Saddler JN (2018) The inhibition of hemicellulosic sugars on cellulose hydrolysis are highly dependant on the cellulase productive binding, processivity, and substrate surface charges. Bioresour Technol 258:79–87. https://doi.org/10.1016/j.biortech.2017.12.006
44. Zhang J, Chu D, Huang J, Yu Z, Dai G, Bao J (2010) Simultaneous saccharification and ethanol fermentation at high corn stover solids loading in a helical stirring bioreactor. Biotechnol Bioeng 105(4):718–728. https://doi.org/10.1002/bit.22593
45. Zhang Z, Wang M, Gao R, Yu X, Chen G (2017) Synergistic effect of thermostable β-glucosidase TN0602 and cellulase on cellulose hydrolysis. 3 Biotech 7(1):54. https://doi.org/10.1007/s13205-017-0672-2
46. Zoghlami A, Paës G (2019) Lignocellulosic biomass: understanding recalcitrance and predicting hydrolysis. Front Chem 7:874

Valorization of Rice Straw for Ethanol Production and Lignin Recovery Through Acidic and Alkaline Pre-treatments

Puneet Pathak, Chhavi Sharma, Sneh Gautam, Shashi Prabha Yadav, and Aparajita Das

Abstract Rice straw is a valuable resource that is rich in beneficial natural substances such as cellulose, hemicellulose, and lignin. These substances are commonly used in the production of biofuels and other value-added products. However, there is still a challenge in applying the biorefinery residue lignin on a large scale. The raw solid products recovered from rice straw usually contain various contaminants and heterogeneous lignin with complex molecular weight distribution. Currently, the most surplus rice straw is inefficiently burned in the field, leading to acute respiratory illnesses and dangerously high pollution levels. This is because of its limited use and low nutritional value as fodder. Therefore, new methods must be developed to use it as a feedstock for producing biofuels and biopower. Recent studies have evaluated changes in the chemical composition, physio-chemical structure, lignin-carbohydrate connections, and accessibility after several pretreatment methods, mainly acid or alkali or combined, to reduce the recalcitrance of rice straw. Utilizing every component of rice straw in a biorefinery-based method makes the process environmentally friendly and sustainable.

Keywords Rice straw · Acid pretreatment · Alkali pretreatment · Ethanol · Lignin · Value-added products

---

P. Pathak (✉) · S. P. Yadav
Research & Development, Agriliv Research Foundation, Sonipat 131001, Haryana, India
e-mail: puneet.pathak@arfmail.in

C. Sharma
University Centre of R&D Biotechnology, Chandigarh University, Mohali 140413, Punjab, India

S. Gautam
Department of Molecular Biology and Genetic Engineering, CBSH, G.G. Pant University of Agriculture and Technology, Pantnagar, India

A. Das
Head Office, Agriliv Research Foundation, New Delhi, India

S. Sundaramurthy et al. (eds.), *Black Liquor Waste Management in the Pulp and Paper Industry*, Sustainable Materials and Technology,
https://doi.org/10.1007/978-981-96-4237-3_8

1 Introduction

The rapid depletion of fossil fuel reserves has spurred interest in the development of alternative energy sources. With the current rate of population growth and energy consumption, the planet is fast approaching its carrying capacity. Hence, there is a pressing need to explore sustainable and innovative methods for producing chemicals, energy, and food. Bioethanol has emerged as a popular biofuel; however, first-generation biofuels have led to a tug-of-war between energy requirements and food security. Second and third-generation biofuels hold the promise of reconciling the demand for energy and food. Researchers worldwide are currently working on the synthesis of ethanol from lignocellulosic biomass, which is known as second-generation bioethanol [93]. The production of second-generation bioethanol using lignocellulosic biomass, which comprises cellulose, hemicellulose, and lignin, offers numerous promising prospects for the biorefinery. As a renewable natural resource that is readily available, the utilization of lignocellulosic biomass as a sustainable feedstock for bioenergy and biochemicals has gained significant attention [2, 12, 20, 35, 72]. Furthermore, using lignocellulosic biomass would not disrupt the food chain supplies, such as with the help of starch-based biomass. In addition to wood, grass, rice straw, bagasse, wheat straw, and several other forest and agricultural plants, lignocellulosic material comprises cellulose, hemicellulose, and lignin.

Effective use of it has the potential to eliminate waste management issues related to forestry and agriculture as well as the usage of fossil fuels [2, 83, 103, 121]. The primary sources of hexose (C6) and pentose (C5) sugars are cellulose and hemicellulose, respectively. These sugars can be fermented to create a range of biofuels, including bioethanol and biobutanol [25] (Fig. 1).

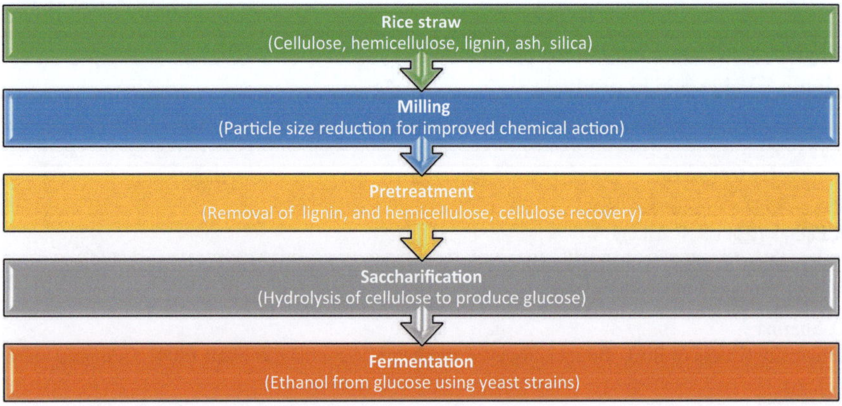

Fig. 1 Basic steps for the production of ethanol from rice straw

2 Rice Straw as a Raw Material for Value-Added Products

Rice is an important cereal crop since it is the primary staple food in most Asian countries. One of the world's most common, inexpensive, and renewable agricultural wastes is rice straw, which is left behind after rice is harvested from the rice field. Reducing resource consumption and environmental degradation can be accomplished through the efficient fractionation and full utilization of straw components [32, 114, 145]. Cellulose, hemicellulose, and lignin components comprise most of the lignocellulosic biomass. Over 20% of the Earth's biosphere's total mass is composed of lignin, which is the second most abundant naturally occurring organic polymer [11]. Rice straw is a waste lignocellulosic residue containing cellulose (28–36%), hemicellulose (23–28%), and lignin (15–23%) [53, 144]. This composition of rice straw shows that it is a suitable lignocellulosic feedstock for the production of second-generation bio-ethanol, a biofuel that can replace gasoline [2, 83, 121]. Moreover, rice straw may be used as a feedstock in the biorefinery concept to produce biobased products and chemicals with additional value. The second-generation biofuel sector will become sustainable and profitable if these biopolymers can be recovered in a single procedure.

3 Pretreatment of Lignocellulosic Raw Material for Delignification

The delignification process results in biomass containing cellulose as a major component. after breaking down the polymeric components (lignin and hemicellulose) of lignocellulosic biomass into simple sugars [45]. Every pretreatment procedure has been used to produce second-generation biofuel wastes and they under utilize at least one of the constituents. It is known that hemicellulose fraction in the form of pentoses and a small amount of lignin can be extracted by diluted acid pretreatment [45, 49], producing biomass that is high in both cellulose and lignin. The most often used physical and chemical pretreatments to break down the barriers and increase cellulose's accessibility to enzymes are diluted acid and alkali treatments [93, 112, 117, 148].

The high value-added chemical production process that uses lignin severely threatens environmental preservation and sustainability. High-quality cellulose production is the primary goal of pretreatment, which is the first step in fractionating lignocellulose in a conventional biorefinery. Pretreatment is energy-intensive. Numerous fractionation techniques have been devised, including deep eutectic solvent, alkaline, dilute acid, organosolv, and SO_2-catalyzed steam explosion [22, 56, 70, 91, 116]. The lignin undergoes structural modification within these processing conditions, making further catalytic valorization of it extremely difficult [87]. As a result, hydrotropic fraction plays a crucial role in identifying a pretreatment technique that is advantageous for exploiting every component of the lignocellulose and has comparatively moderate processing conditions.

4 Detoxification Step Before Fermentation

After acid treatment, a pentose-rich liquid fraction is produced, which can be fermented with the help of organisms that break down pentoses, such as *Pichia pastoris* and *Pichia stipites* [115, 154]. Naturally, furans, furfurals, and phenolics formed during acid treatment restrict fermentation; therefore, these compounds must be eliminated from the fraction before fermentation. Numerous detoxification techniques have been published in the literature [76, 119]. Numerous techniques exist for detoxification, including chemical (neutralisation, calcium hydroxide overlining, activated charcoal treatment, ion exchange resins, and extraction with ethyl acetate), physical (evaporation, membrane-mediated detoxification), biological (enzymatic mediated using laccase, lignin peroxidase), in-situ detoxification, in-situ microbial detoxification, etc. [17].

5 Alkali Treatment

The soda procedure is one of the primary chemical processes used to de-lignify the initial biomass. Soda method is often employed for the delignification of rice straw, sugarcane bagasse, wheat straw, etc. Compared to other procedures, it appears to be the most successful way to break the lignin-carbohydrate connections and ether (C–O–C) bonds between lignin units. The majority of lignin and a fraction of hemicellulose may be solubilized using a variety of alkaline solutions in the soda process, including NaOH, KOH, $CaCO_3$, and $Ca(OH)_2$ [54, 97, 146]. The ether bonds that hold each unit of lignin together break during the soda process, which occurs between 150 and 190 °C. As nucleophiles, the ionized hydroxyl groups target the α- and β-carbons.

A reasonable next step is the delignification of the biomass that remains after acid treatment. The most popular technique for increasing the digestibility of lignocellulosic biomass is alkali pretreatment because diluted alkali treatment causes the biomass to swell, increasing its internal surface area, decreasing its crystallinity, and disrupting the structure of the lignin [117]. The most effective way to disrupt the ester linkages between lignin, hemicellulose, and cellulose is through alkali treatment instead of acid/oxidative reagents [12]. The most often used de-lignifying agent is sodium hydroxide [117].

At a high temperature of 150–190 °C, the ionized hydroxyl groups in the alkaline solution act as nucleophiles and attack the β-carbon or even α-carbon in the lignin structure. Lignin extraction (fragmentation, depolymerization, and solubilization) is dominated by nucleophilic substitution reaction, which has a better effect on lowering the average molecular weight of lignin [36, 54, 143]. Nevertheless, following rigorous chemical processing, the extracted lignin exhibits more molecular weight distribution and chemical structure variation than the original lignin did. The usage of lignin in industry is limited because different molecular weights tend to have varied chemical compositions and functional group contents [31, 69]. Herbaceous crops and agricultural wastes respond well to alkaline chemicals. Alkaline pretreatment can extract lignin from biomass by breaking the ester linkages that bind lignin and xylan together. This process yields higher amounts of cellulose and hemicellulose in the fractions, similar to the kraft or soda pulping process [78].

Applying alkali, typically sodium, potassium, calcium, or ammonium hydroxides to biomass in an ambient environment is known as the alkaline treatment [101]. Alkali pretreatment can be carried out at room temperature for a variety of lengths of time, ranging from seconds to days. The cellulose's internal surface area increases due to the swelling brought on by the NaOH, but the amount of polymerization and crystallinity decrease. The lignin structure breaks as a result. It has been shown that reducing the lignin concentration of hardwood from 24–55% to 20% with NaOH increases the digestibility from 14 to 55%. The efficient removal of lignin from biomass is the main advantage of this technique. NaOH is your greatest option if you're looking for anything that works.

It was discovered to be more effective than alternative techniques at increasing usable interior space. Softwoods with lignin contents more significant than 26%, on the other hand, did not react to diluted NaOH. By adding air or oxygen, the alkaline pretreatment [$NaOH/Ca(OH)_2$], which aids in the removal of lignin, can be rendered more effective. However, the procedure is too costly due to the significant processing costs arising after the initial purchase of lime or other hydroxides. A lot of water is needed to wash the sodium and calcium salts employed in the procedure. Moreover, they are difficult to eliminate [7, 77, 84]. By removing uronic and acetyl from hemicelluloses, the process improves the enzyme accessibility.

Additionally, there is ester bond hydrolysis between xylan and hemicellulose residues. Compared to acid pretreatment, this method reduces the cellulose and hemicellulose solubilization while increasing the cellulose digestibility and lignin solubilization. It has been shown that higher saccharification yields can be obtained by pretreating biomass in an alkaline manner before fermentation. High-temperature alkaline solution pretreatment lowers the cellulose crystallinity, enhances biomass porosity, and breaks down hemicelluloses into sugar monomers, all of which hasten enzymatic hydrolysis. Many caustic salts can be recovered and/or regenerated; alkaline pretreatment is a cost-effective way to decrease sugar degradation. Stimulating solvation and saponification reactions, which cause the cell wall to become porous, sets this method apart. This decreases cellulose, starch crystallinity, and starch polymers while allowing intracellular substances to escape [77, 82, 139, 141, 151].

It increases the rice straw CrI by 10–20% by removing the majority of the hemicelluloses and amorphous lignin from biomass at a temperature that is generally lower than that of DAP, but typically over a longer period of time. According to FTIR measurements, lignin removal and aromatic deformation occurred concurrently with carbohydrate depolymerization. The molecular weight and polydispersity index of rice straw lignin were found to have decreased to 2320 g mol^{-1} and 1.45, respectively, from 2910 g mol^{-1} and 1.87 present in native rice straw following alkali pretreatment. Similarly, in a previous report [48] it was observed that, following alkali pretreatment, lignin's molecular weight decreased and its polydispersity decreased from

3.53 to 1.03. The ^{1}H NMR spectra of the lignin fractions extracted from the native and pretreated rice straw revealed that the cleavage of β-O-4, β-5, and β-β units was indicated by the removal of the representative signals for Hα and Hβ of β-O-4 structures and Hγ of β-5 and β-β structures following NaOH pretreatment [48]. SEM micrographs of the rice straw before and after the alkali pretreatment demonstrated how the natural rice straw's smooth, undamaged, and orderly structure changed into a rough, defibrillated substance [104]. According to another study [34], rice straw inflated and turned into a very porous and accessible substance following alkali pretreatment. Additionally, Khaleghian et al. [60] proposed that, following pretreatment, the packed structure of native rice straw was loosened and the exterior silica layer was removed to increase enzyme accessibility. The alkalinity of the pretreatment determines how much silica is removed. For instance, 1% sodium hydroxide has been shown to remove up to about 75% of the silica in rice straw, but sodium carbonate was only around 30% efficient for the same amount of silica removal [126, 152]. Only under alkaline circumstances does silica from biomass solids move to pretreatment fluid, some of which can precipitate by reducing the pH and then by filtering or centrifuging.

Alkali pretreatment entails applying alkaline solutions, such as NaOH or KOH, to effectively increase the enzyme's accessibility to the cellulose while removing lignin and a portion of the hemicelluloses. The alkali pretreatment can result in a dramatic increase in saccharification yields. Low temperatures can be used for pretreatment, but it will take longer and require a higher base concentration. Alkali treatment seems to be the most efficient technique for cleaving the ester linkages that bind lignin, hemicellulose, and cellulose together and preventing the fragmentation of the hemicellulose polymers when compared to acid or oxidative reagents [38]. The lignin content was reduced by 36% after an alkaline pretreatment of chopped rice straw with 2% NaOH and 20% solid loading at 85 °C for one hour [156]. The exposed and separated microfibrils enhanced the rice straw's external surface area and porosity, which made enzymatic hydrolysis easier. By rupturing the ester bonds that link lignin and xylan, sodium hydroxide pretreatment primarily de-lignifies lignocellulosic biomass, hence enhancing its porosity [132].

To effectively undergo ionization in a high pH environment and increase the solubility of lignin, a nucleophilic substitution process was performed on the phenolic hydroxyl groups on the side chains of lignin units [36, 143]. It was reported in recent years to treat biomass to improve the solubilization of lignin by adding an organic solvent, such as ethanol [28, 46, 67, 81], diethylene glycol [43], ethanolamine [108, 109] or dimethyl formamide [39], as a co-solvent to the Soda process. A high-boiling-point solvent may be utilized at a higher processing temperature without significant pressure. However, low-boiling-point solvents are often preferred for their ease of recovery and relative affordability.

6 Concentrated or Diluted Acid

Acid pretreatment is the most commonly employed procedure due to its high efficiency. Acid pretreatment increases the enzyme accessibility of cellulose and solubilizes hemicellulose and lignin. Concentrated acid generates inhibitory compounds. However, diluted acid can still be used for acid pretreatment. To improve glucose recovery, acids hydrolyze cellulose of its hemicellulose and lignin. In full scale systems, equipment that is resistant to corrosion is required because concentrated acids are toxic, corrosive, and dangerous. Large volumes of breakdown products are produced, though, when these compounds are utilized at high temperatures. Among other things, they can damage the environment, corrode machinery, and pollute leftovers. Elevated concentrations of these substances shorten reaction times and do not require enzymes; low concentrations necessitate increased pressure and temperature for the best hydrolysis efficiency. The requirement to neutralize the samples after they have been treated is one of the disadvantages of utilizing concentrated chemicals, which increases the process's overall cost [77, 84, 139]. Diluted acid pretreatment is the greatest option when used on an industrial basis. Dilute acid pretreatment can be done in one of two ways: either for a brief period of time at a high temperature (>180 °C) or for a longer period of time (30–90 min) at a lower temperature (>120 °C). Distilled acid appears to be a preferable treatment for lignocellulosic biomass prior to industrial usage. For a broad variety of lignocellulosic biomass, this has been explored. Fumaric and maleic acids are examples of organic acids that may improve cellulose hydrolysis for the manufacture of ethanol [7, 77, 105, 139]. High hydrolysis yields have been observed with diluted sulfuric acid, which is also the most widely used acid. Investigations have been conducted on H_2SO_4, H_3PO_4, NH_3, $C_2H_4O_3$, $C_2H_2O_4$, HCOOH, CH_3COOH, and $C_4H_4O_4$ [77].

Fourier Transform Infrared Spectroscopy (FTIR) analysis showed that peak intensities for hemicelluloses were either decreased or lessened following DAP. Nevertheless, the region containing crystalline cellulose did not alter. Teramura et al. [133] discovered that following DAP of rice straw, there was a significant decrease in ferulate NMR signals, which are indicative of lignin-carbohydrate linkage and other aliphatic linkages such as 5–5/4-O-β (dibenzodioxin), β-O-4 (β-aryl ether), β-O-4-H/G (β-aryl ether p-hydroxyphenyl/guaiacyl), β-O-4-S (β-aryl ether syringyl), and β-5 (phenylcoumaran). The breakdown of cross-bridges between lignin and carbohydrate was further confirmed by the decrease in hemicellulose and arabinofuranoside NMR signals. Meng et al. [80] discovered that, following DAP, the smooth, intact physical structure of native rice straw was changed to a wart/papillae-type structure by the breakdown of cuticle wax and silica layers. Additionally, Sun et al. [127] stated that following DAP, fractures and divided fibrous structures were seen. The elimination of hemicellulose and the opening of macropores, which were roughly twice as large as those in the untreated rice straw, were proposed as the causes of this impact. A large portion of the silica in pretreated solids has reportedly been retained by DAP [92].

Anaerobic digestibility of lignocellulose is improved by pretreating it with acids at room temperature. Hemicellulose is primarily affected by diluted acid pretreatment, but lignin breakdown is not significantly affected. Acid pretreatment will solubilize the hemicellulose, improving enzyme accessibility to the cellulose. Mineral acids such as HCl and H_2SO_4 are typically used for acid pretreatment. The residual carbohydrates in the processed biomass must be hydrolyzed by the enzyme cellulase after treatment with diluted acid. A straightforward one-stage procedure known as "diluted acid pretreatment" involves treating the biomass with diluted sulfuric acid for a predetermined amount of time at the right temperature and acid concentration. A two-stage procedure was created at the National Renewable Energy Laboratory (NREL) in Golden, Colorado (USA), to lower the amount of enzymes needed. The lack of lignin removal and low sugar yield of the method has limited the amount of literature available on the hydrolysis of rice straw using diluted acid [125].

Pretreated rice straw had a rough and porous surface with recognizable micropores. The rougher surface and a higher surface area resulting from eliminating hemicelluloses are often connected to improved enzymatic hydrolysis. Due to sulfuric acid's low cost and versatility, diluted acid pretreatment, or DAP, is one of the most widely used pretreatment techniques. The elimination of lignin and amorphous hemicellulose, followed by enrichment in the concentration of crystalline cellulose, may be the cause of the observed rise in the crystallinity Index (CrI) of rice straw following DAP, according to X-ray diffraction (XRD) study [4, 5, 52].

7 Organic Solvents

Selecting organic solvents is an additional technique. Dispersion forces, permanent dipole–dipole force, and hydrogen bonds all significantly impact the soluble lignin in solvents. According to Hansen's solubility parameter theory (HSP theory) [86], methanol [111], ethanol [28, 40], acetone [14], tetrahydrofuran (THF) [79], ethylene glycol [6], acetic acid [61], and so on, are examples of frequently used organic solvents. Subsequent extraction using organic solvents can effectively separate lignin molecules in particular molecular weight ranges because lignin has distinct physical and chemical characteristics at different molecular weights [33, 68]. Regarding accessible solvents, low-molecular-weight lignin is generally more diverse than high-molecular-weight lignin [98, 107]. This technique produces high-purity lignin while being safe for the environment and enabling effective separation. However, there are also drawbacks, such as the expense of organic solvents, concerns about safety, and challenges with solvent recovery.

Rice straw can be used in thermochemical and biochemical processes to produce biopower and biofuels. In order to decrease rice straw recalcitrance for the biological platform, a pretreatment step is required. This can be achieved by changing the way that cellulose, hemicellulose, and lignin interact, increasing cellulose accessibility, removing lignin-carbohydrate complexes, or reducing the cellulose crystallinity. These effects raise the yields from biomass's enzymatic hydrolysis, which raises the yields from ethanol fermentation.

8 Combined Pretreatment

According to Kun et al. [64], rice straw was pretreated with alkali with the help of photocatalysis, which effectively altered the microstructure and physical characteristics of the straw. It also decreased the amount of lignin in the straw, which increased the rate at which the straw was hydrolyzed by the enzymes. While the second stage of treatment with alkaline peroxide enhanced the dissolution of larger molecular size hemicelluloses that were rich in xylose, the alkali treatment of rice straw in the absence of H_2O_2 favored solubilization of the small molecular size hemicelluloses, which are rich in glucose and likely originating from o-glucan [126]. When used in conjunction with other techniques, microwave pretreatment is becoming a significant and effective technology. According to Zhu et al. [159], there are several combinations of microwave pretreatment and acid or alkali that removes lignin and hemicellulose, respectively, from rice straw. The microwave pretreatment eliminates more lignin than alkali pretreatment alone. The findings demonstrate that, for the same energy consumption, the effects of greater microwave power with a shorter pretreatment period and lower microwave power with a longer pretreatment duration were nearly identical in terms of weight reduction and composition. Certain pretreatment reactions are improved by the microwave, yet the precise mechanism is yet unknown. Using an electron beam accelerator, Xin and Kumakura [149] demonstrated radiation pretreatment of rice straw in the presence of NaOH solutions. The lignocellulosic structure is altered by electron beam irradiation, making it easier for NaOH solution to penetrate the lignocellulosic complex and speed up the reaction, which leads to the easier removal of lignin and a slight degradation of cellulose or hemicellulose that was exposed to radiation. This increases the accessibility of enzymes.

Following a sequential pretreatment with 3% *v/v* of sulfuric acid and 4% *w/v* of sodium hydroxide, rice straw had approximately 90% of its hemicelluloses and 55% of its lignin removed. *Saccharomyces cerevisiae* HAU was used to ferment the 787 mg/g of reducing sugars produced by saccharifying the pretreated rice straw using cellulases. The post-acid treatment acid hydrolysate was additionally fermented using *Pichia stipitis* NCIM 3499. The fermentation of hexose and pentose yields 26.9 and 9.4 g/L of ethanol, respectively (79 and 93% fermentation efficiency, respectively) [59]. After the wasted liquor was recovered from the alkali treatment, high-purity (98%) lignin (146 mg/g dry substrate) was precipitated by direct acidification with HCl. FTIR and TGA were used to characterize the extracted lignin. The study's overall goal was to efficiently turn rice straw's primary polymers into valuable products that would increase the sustainability of the bioethanol production process [59].

Research has been conducted on the use of alkali and acidic pretreatment methods for recovering glucan and xylan from rice straw (Table 1).

9 Saccharification

The second stage in ethanol synthesis from lignocellulosic materials is enzymatic hydrolysis. Enzymes are used to cleave the cellulose and hemicellulose polymers. Hemicellulose comprises polymers of several sugars, including mannan, xylan, glucan, galactan, and arabinan, while cellulose often only contains glucans. Thus, glucose is the primary hydrolysis product of cellulose, while numerous pentoses and hexoses are produced by hemicellulose [129]. On the other hand, a high lignin content inhibits the end product, decreases the pace and yield of hydrolysis, and blocks the accessibility of enzymes. Apart from lignin, cellobiose and glucose also function as potent cellulase inhibitors [62].

The black liquor obtained after alkaline pretreatment is similar to kraft lignin and can be selectively precipitated to obtain pure lignin fractions. The biomass left after alkali treatment is rich in cellulose, which can be hydrolyzed using cellulases to simpler sugars, i.e., glucose, which can further be fermented to ethanol by hexose fermenting organisms like *Saccharomyces* and *Zymomonas mobilis*, etc. [66, 130].

Particle size, liquid to solid ratio, type and concentration of acid used, temperature, and reaction time are some of the factors that affect the yields of lignocellulose to monomeric sugars and byproducts. Other factors include the length of macromolecules, degree of cellulose polymerization, configuration of the cellulose chain, and association of cellulose with other protective polymeric structures within the plant cell wall, such as lignin, pectin, hemicellulose, proteins, and mineral elements. Recent developments in enzyme technology for the conversion of cellulosic biomass to sugars have brought tremendous success in lignocellulosic ethanol research. When enzymatic hydrolysis is used to hydrolyze hemicellulose, gentle conditions are often met, such as low pressure and extended retention time. Valdes and Planes conducted research on the hydrolysis of rice straw at 80–100 °C using 5–10% H_2SO_4. The highest sugar production was recorded at 100 °C for 240 min using 10% H_2SO_4 [137]. Yin et al. [153] successfully hydrolyzed over 70% of the pentoses in the hemicellulose portion of rice straw at 110–120 °C using 2% H_2SO_4. When rice straw was hydrolyzed with several acids at varied concentrations (0.5–1% H_2SO_4, 2–3% HCl, and 0.5–1% H_3PO_4), previous studies [138, 153] have discovered that the pentosans in the straw transformed into a monosaccharide solution that could be fermented after three hours of retention [138]. Roberto et al. [106] investigated, in a 350-L batch reactor, the effects of H_2SO_4 concentration and retention time on the yield of sugars and byproducts from rice straw at a relatively low temperature (121°C) and for a considerable amount of time (10–30 min). It was discovered that a high yield of xylose could be achieved with an acid concentration of 1% and a retention period of 27 min (77%). During the 48-h enzymatic hydrolysis, the pretreatment of the straw with diluted sulfuric acid produced a sugar yield of 0.72 g g^{-1}, which

Table 1 Studies on alkali and acidic pretreatment of rice straw for glucan and xylan recovery

Raw material	Pretreatment type	Pretreatment conditions	Effects	Glucan/xylan recovery, %	References
Rice straw	Alkali	Soild to liquid ratio 10% (*w/v*), 1–5% (*w/v*) NaOH, using incubator shaker (150 rpm, 40 °C for 1 h), and autoclave-sterilized at 121 °C for 30 min	47% reduction of lignin and about 50% increase in glucan	57.78/11.84	[88]
		10% (*v/v*) ammonia solution, solid to liquid ratio 1:10 at 100 °C for 6 h	Glucan and xylan recovery rate is quite high	84.0/52.0	[85]
		Lime treatment, solid and liquid ratio 1:20, 120 °C. 1 h	No need to separate the solid–liquid and washing steps	95.2/81.5	[96]
		Soild to liquid ratio 10% (*w/v*), 1–5% (*w/v*) NaOH, using incubator shaker (150 rpm, 40 °C for 1 h), and autoclave-sterilized at 121 °C for 30 min	47% reduction of lignin and about 50%.increase in glucan	57.78/11.84	[88]
	Microwave alkali	Soild to liquid ratio 8% (*w/v*), 1% NaOH, boiling temperature, 30 min, microwave frequency: 2450 MHz), 700 W for 30 min	76% reduction of lignin and reduced crystallinity	69.2/10.2	[160]
		Microwave frequency: 2450 MHz; Power 700, 500, and 300 W, respectively; 1% NaOH; 1:8 solid and liquid ratio; 15 min to 2 h	Higher removal of hemicellulose and lignin helps the pretreated material's accessibility towards hydrolytic enzymes	69.2/10.2	[160]

(continued)

Table 1 (continued)

Raw material	Pretreatment type	Pretreatment conditions	Effects	Glucan/ xylan recovery, %	References
	Acid	Soaking in the 0.5% sulfuric acid; 1:6.66 solid to liquid ratio; 20 h; In the reactor: steam-heating 1.5 min, 15 bar pressure, residence time 10 min	99% hemicelluloses removal helps enzymatic hydrolysis	55/1%	[1]
		Soild to liquid ratio 10% (*w/v*) dilute acid hydrolysis with 2% (*w/v*) sulfuric acid, 132 °C for 30 min under pressure	Better enzymatic hydrolysis due to reduced crystallinity of the rice straw, xylan removal	50.7/10.8	[89]
		Sulfuric acid 1% (*w/w*), 1:10 ratio, acid reaction time, 1–5 min at 160 °C or 180 °C for 1 min residence time	>90% solubilization of xylan	57/4.7	[49]
	Organosolv	Solid-to-liquid ratio 1:8 of 75% (*v/v*), aqueous 1% (*w/w*) ethanol containing sulfuric acid as a catalyst. using high-pressure stainless steel vessel at 180 °C for 60 min	Solvent recovery	62.3/22.4	[8]
	Organosolv with microwave	Acetic acid, microwave: 100–700 W, 2–5 min	51.54% removal ratio of lignin	–	[42]

was more than the output of 0.46 and 0.60 g g^1 from steam pretreatment [1]. When the substrate concentration was raised from 20 to 50 and 100 g L^{-1}, respectively, the sugar production decreased to 13 and 16%. Aderemi et al. [3] investigated the kinetics of *Aspergillus niger's* for the production of glucose from rice straw. The rice straw particle size was found to reduce from 425 to 75 lm, resulting in an increase in glucose output from 43 to 87%. The ideal temperature and pH range were found to be 45–50 °C and 4.5–5.0, respectively. The results demonstrate that cell loading, substrate concentration, and rice straw pretreatment all affect the amount and pace of glucose production. According to Kun et al. [64], the hydrolysis rate of rice straw produced by the enzymatic hydrolysis of alkali-enhanced photocatalysis was 2.56 times higher than that of the alkali process. On the other hand, rice straw treated with ammonia increased the amount of monomeric sugars from 11% in the untreated to 61% [124]. Combining enzymes like cellulase, xylanases, and pectinases instead of only cellulase boosts the hydrolysis efficiency of lignocellulosic biomass [157]. However, although this procedure is highly desirable from an ecological standpoint, it is very expensive.

10 Fermentation

Either separate enzymatic hydrolysis and fermentation (SHF) or simultaneous saccharification and fermentation (SSF) methods can be used to convert the cellulose and hemicellulose components of rice straw into ethanol. SSF is preferred more due to its relatively cheap prospective costs [147]. Because it minimizes product inhibition, the ethanol yield is higher than that of SHF. The discrepancy in the ideal temperature of the microorganisms that ferment and the hydrolyzing enzymes is one of the disadvantages of this procedure. The majority of publications indicate that 40–50 °C is the ideal temperature for enzymatic hydrolysis; nevertheless, bacteria that produce large amounts of ethanol typically cannot withstand this high temperature. By using thermotolerant microorganisms like *Zymomonas mobilis*, *Candida lusitaniae*, and *Kluyveromyces marxianus*, or by cultivating a mixed culture of some microbes like *Brettanomyces clausenii* and *Saccharomyces cerevisiae*, this issuc can be avoided [41, 120]. *Pachysolen tannophilus* and *Candida brassicae* were used in Punnapayak and Emert's [102] study on the solid-state fermentation (SSF) of alkali-pretreated rice straw. Throughout the experiment, *P. tannophilus* produced higher ethanol yields than *C. brassicae*. They did, however, only come close to achieving the theoretical ethanol production of less than 30%. Using *Rhizopus oryzae*, *S. cerevisiae*, and *Mucor indicus* in SSF of acid-pretreated rice straw produced an overall output of 40–74% of the maximum theoretical ethanol production [58]. Zhu et al. [159] investigated the SSF of microwave/alkali and alkali-pretreated rice straws to ethanol using cellulase from *T. reesei* and *S. cerevisiae*. The ideal conditions resulted in a 29.1 g L^{-1} ethanol concentration and a 61.3% ethanol yield. The study demonstrates that compared to rice straw pre-treated by alkali alone, ethanol synthesis from microwave/alkali pre-treated straw had lower enzyme loading, a shorter reaction time, and a greater ethanol concentration and yield. Numerous reports claim that when producing ethanol from rice straw, simultaneous saccharification and fermentation (SSF) is preferable to traditional saccharification and subsequent fermentation because the SSF process can increase ethanol yields by removing end-product inhibition of the saccharification process and get rid of the requirement for separate reactors for fermentation and saccharification [16]. Abedinifar et al. [1] investigated the distinct enzymatic hydrolysis and fermentation of rice straw by *M. indicus*, *R. oryzae*, and *S. cerevisiae*. According to their research, *M. indicus* can synthesis ethanol from pentoses. This species appears to be a good strain for lignocelluloses (rice straw in particular) to produce ethanol. Apart from SSF and SHF, there exists an additional procedure known as consolidated bioprocessing (CBP). This technique involves the simultaneous operation of ethanol fermentation, biomass hydrolysis, and cellulase generation in a single reactor. The best bacterium for this process will be one like *Clostridium phytofermentans*, which is capable of fermenting cellulose effectively and producing ethanol directly. The two main sugars in the lignocellulosic hydrolysates are glucose and xylose. The primary challenge in co-fermenting these two sugars with two microorganisms is the incapacity to concurrently provide the two strains with ideal environmental conditions [18]. The majority of prior research on strain co-cultures found that, although xylose fermentation was frequently slow and inefficient owing to the two strains' competing oxygen requirements and/or the catabolite repression on the xylose assimilation brought on by the glucose, the fermentation of glucose in the sugar mixture proceeded efficiently with a traditional glucose-fermenting strain [44, 63]. Methods from the fields of strain and process engineering have been used to get around these issues and boost system performance. Process engineering examples include immobilisation of one of the strains [44], co-immobilization fermentation in two bioreactors, and continuous culture [30, 44, 65, 131]. two stages of fermentation in a single bioreactor (i.e., sequential culture) [10, 44], and separate fermentation in two bioreactors [44, 131].

11 Lignin Recovery

After cellulose, lignin is the second-largest source of biomass in the world. It is a three-dimensional amorphous polymer and one of the three main components of lignocellulose, which include adhesion, biodegradability, thermal stability, and antibacterial activity [12, 35]. Lignin is a multilayered, three-dimensional polyaromatic matrix that blocks enzymes from penetrating cellulose [47]. Lignin, along with other process waste products, is a primary byproduct of bioethanol synthesis and is typically disposed of in water bodies. This trash contains several harmful lignin-derived compounds, making the process hazardous to the environment. But, if the lignin from this waste could be used, this issue may be resolved. Furthermore, lignin has the potential to yield several valuable products, including fuel, phenolic resins, emulsifiers, dispersants, chelating agents, fat mimetics for emulsion-based foods, bioplastics, nanocomposites, and nano-particles, as adsorbents in solutions, alternative to phenol–formaldehyde, protective UV- absorbents, and dispersants [13, 23, 95, 122, 135, 136, 150]. The process's intermediates can be used to create biopolymers of commercial value, including high-value carbon fibers, dispersants, emulsifiers, adhesives, and bio-polyester fibers.

Hemicellulose and cellulose in rice straw biorefineries break down into sugars for further biochemical transformations, leaving lignin to be recovered for further value addition [21, 103, 123]. Effective lignin recovery emerged as a critical concern in the overall design of the biorefinery processes, and the use of lignin will increase the overall economics of the biorefinery processes.

Most cellulose and hemicellulose components are hydrolyzed into soluble sugar solutions, leaving behind the brown solid (BS) residue, an insoluble waste byproduct that remains after the acid or enzymatic hydrolysis process having high lignin content. One significant source of lignin that can be obtained in good quality is lignin recovery from the brown solid residue [113, 155]. Researchers have developed the process for lignin extraction from brown solids to increase the efficiency of biomass utilization in lignocellulosic biorefinery processes concentrated on the recovery of lignin directly from biomass [29, 118].

Results of carbon reduction that replace fossil fuels will have larger marginal advantages [2, 9, 37, 142]. Nonetheless, lignin's molecular weight distribution and chemical structure are highly varied and complex. In reality, only 1–2% of lignin is utilized by the industry [37]. In light of this, completely reducing the variability of extracted lignin has become a crucial concern. Natural lignin is an irregularly cross-linked polyphenolic polymer in terms of molecular structure. It is mainly made up of three monomers, which in the structure of lignin are p-hydroxyphenyl (H), guaiacyl (G), and syringyl (S). These monomers are also known as coniferyl alcohol, sinapyl alcohol, and p-coumaryl alcohol. Be aware that different species may have varying ratios of monomer makeup. Either C–O–C (β-O-4, α-O-4, or 4-O-5) or C–C (β-1, β-5, β-β, or 5–5) links bind the polyphenol polymers together. Of these linkages, β-O-4 linkage makes up the most significant percentage, ranging from 50 to 65% [19, 27, 74, 158].

Therefore, the lignin fractionation process must be utilized to obtain a stable lignin product with a certain molecular weight and chemical properties. Previous research have identified three common strategies for lignin fractionation.

Treatment with acid precipitation is the first technique. Acids like HCl [55], H_2SO_4 [100, 128] are frequently employed to reduce the pH of lignin solutions. The negatively charged lignin colloid particles are neutralized as a result. At this point, lignin clumps and precipitates due to decreasing repulsive interactions between its molecules [110]. Low-molecular-weight lignin requires a lower pH to bind because it has a higher surface negative charge. On the other hand, lignin with a high molecular weight can be gathered at a higher pH [99]. Consequently, it is possible to efficiently separate lignin with varying molecular weights by using successive pH control [15, 73, 140]. The benefits of this approach are minimal energy use, quick response times, and straightforward operation. However, a considerable amount of lignin-carbohydrate complexes may be generated, the recovered lignin's purity is low, and the expense of acid recovery is substantial [134].

The insoluble brown solid (BS) residue that is left over after acidic or alkaline hydrolysis during the manufacture of bioethanol was utilized as the starting point for separation and purification [71]. A considerable amount of lignin with a high degree of diversity and heterogeneous molecular weight distribution, together with a number of contaminants and organic materials, are still present in BS even after the majority of the cellulose and hemicellulose have been removed. A very high level of purity (>90%) can be achieved by treating the lignin with an acid precipitation treatment using the soda-EG procedure. More significantly, over 90% of the lignin is recovered.

The lignin solution is more precisely segregated into lignin with a high (>105 g/mol) and low (<104 g/mol) molecular weight range by dead-end and cross-flow systems using commercially available membranes with specified molecular weight cut-off (MWCO). High-molecular-weight lignin is frequently utilized in polyurethane foam, adhesives, absorbents, and as a starting point for the production of carbon fibers. Conversely, low-molecular-weight lignin can be utilized to create bioplastics, fertilizers, and natural coagulants for water treatment [2, 9, 142]. In order to greatly increase the economic value of biorefinery technology, the ultimate goal should be to isolate a final solid product with high purity and a particular molecular weight range of lignin that may be used for distinct high-value-added downstream applications.

Membrane separation is a simple technique that can be used constantly without requiring the addition of chemicals. It distinguishes between lignin molecules with different molecular weights by using membranes with distinct pore diameters. Moreover, all lignin molecules have molecular weights between 100,000 and 1,000 kDa, commensurate with the ultrafiltration operating range (UF). Ceramic membranes like TiO_2 coated with ZrO_2 [24, 90] and polymer membranes like polyvinylpyrrolidone [57], polysulfone [75], and poly(ether sulfone) [51] are examples of common materials.

The filtrate can also be classified into two categories based on the direction in which it flows: dead-end filtration and cross-flow filtration. Membrane separation offers superior molecular weight distribution and greater quality extracted lignin due to its ability to remove carbohydrates and inorganic components from the lignin solution more thoroughly than the other two processes. In addition, the extracted lignin also has improved thermal characteristics due of the reduced inorganic elements in its composition [50, 94]. Despite the unresolved issue of surface fouling on the membrane with increased usage [26], using membranes for lignin fractionation is still a well-researched area of study in recent years.

12 Conclusions

The production of bioethanol using lignocellulosic biomass requires an efficient and sustainable production system that is both cost-effective and environmentally friendly. Rice straw is a promising candidate for bioethanol synthesis due to its attractive composition and wide availability, particularly in the context of the growing demand for second-generation biofuels. The most desirable method for the conversion of rice straw into fermentable sugars is the use of hydrolyzing enzymes, which is also considered environmentally friendly. However, the development of an effective pretreatment technique to remove unwanted portions and obtain readily available sugars is a necessary step for the process. Although significant progress has been made in this regard, the high lignin and ash content of rice straw presents several obstacles to developing economically feasible technology. Researchers are currently focused on using rice straw as a raw material for synthesizing bioethanol and recovering lignin, rather than establishing a self-sustaining lignocellulosic biorefinery. Diluted acid and alkali pretreatments can be used to recover each fraction of lignocellulose. However, to achieve a more sustainable biorefinery-based approach, all biopolymers of lignocellulose should be retrieved in a single step. This would enable the total value of the feedstock to be obtained and aid in the production of many products.

References

1. Abedinifar S, Karimi K, Khanahmadi M, Taherzadeh MJ (2009) Ethanol production by *Mucor indicus* and *Rhizopus oryzae* from rice straw by separate hydrolysis and fermentation. Biomass Bioenergy 33(5):828–833
2. Abraham A, Mathew AK, Sindhu R, Pandey A, Binod P (2016) Potential of rice straw for bio-refining: an overview. Bioresour Technol 215:29–36
3. Aderemi BO, Abu E, Highina BK (2008) The kinetics of glucose production from rice straw by *Aspergillus niger*. African J Biotechnol 7(11), 1745–1752
4. Agrawal R, Gaur R, Mathur A, Kumar R, Gupta RP, Tuli DK, Satlewal A (2015) Improved saccharification of pilot-scale acid pretreated wheat straw by exploiting the synergistic behavior of lignocellulose degrading enzymes. RSC Adv 5(87), 71462–71471
5. Agrawal R, Satlewal A, Kapoor M, Mondal S, Basu B (2017) Investigating the enzyme-lignin binding with surfactants for improved saccharification of pilot scale pretreated wheat straw. Bioresour Technol 224:411–418
6. Alriols MG, Tejado A, Blanco MA, Mondragon I, Labidi J (2009) Agricultural palm oil tree residues as raw material for cellulose, lignin and hemicelluloses production by ethylene glycol pulping process. Chem Eng J 148(1):106–114
7. Alvira P, Tomás-Pejó E, Ballesteros M, Negro MJ (2010) Pretreatment technologies for an efficient bioethanol production process based on enzymatic hydrolysis: a review. Bioresour Technol 101(13):4851–4861
8. Amiri H, Karimi K, Zilouei H (2014) Organosolv pretreatment of rice straw for efficient acetone, butanol, and ethanol production. Bioresour Technol 152:450–456
9. Bajwa DS, Pourhashem G, Ullah AH, Bajwa SG (2019) A concise review of current lignin production, applications, products and their environmental impact. Ind Crops Prod 139:111526
10. de Bari I, Cuna D, Nanna F, Braccio G (2004) Ethanol production in immobilized cell bioreactors from mixed sugar syrups and enzymatic hydrolysates of steam exploded biomass. Appl Biochem Biotechnol 114:539–557
11. Behling R, Valange S, Chatel G (2016) Heterogeneous catalytic oxidation for lignin valorization into valuable chemicals: what results? What limitations? What trends? Green Chem 18(7):1839–1854
12. Binod P, Sindhu R, Singhania RR, Vikram S, Devi L, Nagalakshmi S, Kurien N, Sukumaran RK, Pandey A (2010) Bioethanol production from rice straw: an overview. Bioresour Technol 101(13):4767–4774
13. Boeriu CG, Bravo D, Gosselink RJ, van Dam JE (2004) Characterisation of structure-dependent functional properties of lignin with infrared spectroscopy. Ind Crops Prod 20(2):205–218
14. Boeriu CG, Fiţigău FI, Gosselink RJ, Frissen AE, Stoutjesdijk J, Peter F (2014) Fractionation of five technical lignins by selective extraction in green solvents and characterisation of isolated fractions. Ind Crops Prod 62:481–490

15. Brenelli LB, Mandelli F, Mercadante AZ, de Moraes Rocha GJ, Rocco SA, Craievich AF, Goncalves AR, da Cruz CD, de Oliveira NM, Squina FM (2016) Acidification treatment of lignin from sugarcane bagasse results in fractions of reduced polydispersity and high free-radical scavenging capacity. Ind Crops Prod 83:94–103
16. Chadha BS, Kanwar SS, Saini HS, Garcha HS (1995) Hybrid process for ethanol production from rice straw. Acta Microbiol Immunol Hung 42(1):53–59
17. Chandel AK, da Silva SS, Singh OV (2011) Detoxification of lignocellulosic hydrolysates for improved bioethanol production. Biofuel Prod-Recent Dev Prospect 10:225
18. Chandrakant P, Bisaria VS (1998) Simultaneous bioconversion of cellulose and hemicellulose to ethanol. Crit Rev Biotechnol 18(4):295–331
19. Chatel G, Rogers RD (2013) Review: oxidation of lignin using ionic liquids – an innovative strategy to produce renewable chemicals. ACS Sustain Chem Eng 322–339. https://doi.org/10.1021/sc4004086
20. Chatel G, Rogers RD (2014) Oxidation of lignin using ionic liquids-an innovative strategy to produce renewable chemicals. ACS Sustain Chem Eng 2(3):322–339
21. Chen WH, Hsu MH, Wu AY, Hwang WS (2017) Efficient extraction and recovery of xylan and lignin from rice straw using a flow-through hydrothermal system. J Taiwan Inst Chem Eng 79:103–109
22. Cheng J, Leu SY, Zhu JY, Jeffries TW (2014) Ethanol production from non-detoxified whole slurry of sulfite-pretreated empty fruit bunches at a low cellulase loading. Bioresour Technol 164:331–337
23. Chung BY, Lee JT, Bai HW, Kim UJ, Bae HJ, Wi SG, Cho JY (2013) Enhanced enzymatic hydrolysis of poplar bark by combined use of gamma ray and dilute acid for bioethanol production. Radiat Phys Chem 81(8):1003–1007
24. Costa CA, Pinto PC, Rodrigues AE (2018) Lignin fractionation from *E. Globulus* kraft liquor by ultrafiltration in a three stage membrane sequence. Sep Purif Technol 192:140–151
25. Cotana F, Cavalaglio G, Nicolini A, Gelosia M, Coccia V, Petrozzi A, Brinchi L (2014) Lignin as co-product of second generation bioethanol production from ligno-cellulosic biomass. Energy Procedia 45:52–60
26. Courbalay M, Villain-Gambier M, Klem A, Ziegler-Devin I, Dumarcay S, Trebouet D (2021) Influence of pH and fouling characterization during membrane process for lignin recovery from the process water of thermomechanical pulping. Sep Purif Technol 275:119162
27. Crestini C, Lange H, Sette M, Argyropoulos DS (2017) On the structure of softwood kraft lignin. Green Chem 19(17):4104–4121
28. Dagnino EP, Felissia FE, Chamorro E, Area MC (2017) Optimization of the soda-ethanol delignification stage for a rice husk biorefinery. Ind Crops Prod 97:156–165
29. Davis KM, Rover M, Brown RC, Bai X, Wen Z, Jarboe LR (2016) Recovery and utilization of lignin monomers as part of the biorefinery approach. Energies 9:808
30. Delgenes, JP, Laplace JM, Moletta R, Navarro JM (1996) Comparative study of separated fermentations and cofermentation processes to produce ethanol from hardwood derived hydrolysates. Biomass Bioenergy 11:353–360
31. Dorrestijn E, Laarhoven LJ, Arends IW, Mulder P (2000) The occurrence and reactivity of phenoxyl linkages in lignin and low rank coal. J Anal Appl Pyrol 54(1–2):153–192
32. Dusselier M, Mascal M, Sels BF (2014) Top chemical opportunities from carbohydrate biomass: a chemist's view of the biorefinery. In: Selective catalysis for renewable feedstocks and chemicals, pp 1–40
33. Duval A, Vilaplana F, Crestini C, Lawoko M (2016) Solvent screening for the fractionation of industrial kraft lignin. Holzforschung 70(1):11–20
34. Elumalai S, Agarwal B, Sangwan RS (2016) Thermo-chemical pretreatment of rice straw for further processing for levulinic acid production. Bioresour Technol 218:232–246
35. Figueiredo P, Lintinen K, Hirvonen JT, Kostiainen MA, Santos HA (2018) Properties and chemical modifications of lignin: towards lignin-based nanomaterials for biomedical applications. Prog Mater Sci 93:233–269.

36. Galkin MV, Samec JS (2016) Lignin valorization through catalytic lignocellulose fractionation: a fundamental platform for the future biorefinery. ChemSusChem 9(13):1544–1558
37. Garlapati VK, Chandel AK, Kumar SJ, Sharma S, Sevda S, Ingle AP, Pant D (2020) Circular economy aspects of lignin: towards a lignocellulose biorefinery. Renew Sustain Energy Rev 130:109977
38. Gaspar M, Kalman G, Reczey K (2007) Corn fibre as a raw material for hemicellulose and ethanol production. Process Biochem 42:1135–1139
39. Ghozatloo A, Mohammadi-Rovshandeh J (2007) Organosolv pulping of rice straw by dimethylformamide and optimization of handsheet properties. Cellul Chem Technol 41:175–178
40. Goldmann WM, Ahola J, Mikola M, Tanskanen J (2019) Solubility and fractionation of Indulin AT kraft lignin in ethanol-water media. Sep Purif Technol 209:826–832
41. Golias H, Dumsday GJ, Stanley GA, Pamment, NB (2002) Evaluation of a recombinant *Klebsiella oxytoca* strain for ethanol production from cellulose by simultaneous saccharification and fermentation: comparison with native cellobiose-utilising yeast strains and performance in co-culture with thermotolerant yeast and *Zymomonas mobilis*. J Biotechnol 96(2):155–168
42. Gong G, Liu D, Huang Y (2010) Microwave-assisted organic acid pretreatment for enzymatic hydrolysis of rice straw. Biosyst Eng 107(2):67–73
43. Gonzalez M, Canton L, Rodriguez A, Labidi J (2008) Effect of organosolv and soda pulping processes on the metals content of nonwoody pulps. Bioresour Technol 99:6621–6625
44. Grootjen DRJ, Jansen ML, Van der Lans RGJM, Luyben KCA (1991) Reactors in series for the complete conversion of glucose/xylose mixtures by *Pichia stipitis* and *Saccharomyces cerevisiae*. Enzyme Microb Technol 13(10):828–833
45. Gupta R, Sharma KK, Kuhad RC (2009) Separate hydrolysis and fermentation of Prosopisjuliflora, a woody substrate, for production of cellulosic ethanol by *Saccharomyces cerevisae and Pichiastipitis-NCIM* 3498. Bioresour Technol 100:1214–1220
46. Guvenatam B, Heeres EHJ, Pidko EA, Hensen EJM (2016) Lewis acid-catalyzed depo lymerization of soda lignin in supercritical ethanol/water mixtures. Catal Today 269:9–20
47. Han KJ, Pitman WD, Kim M, Day DF, Alison MW, McCormick ME, Aita G (2012) Ethanol production potential of sweet sorghum assessed using forage fiber analysis procedures. Glob Change Biol Bioenergy 5:358–366
48. He Y, Pang Y, Liu Y, Li X, Wang K (2008) Physicochemical characterization of rice straw pretreated with sodium hydroxide in the solid state for enhancing biogas production. Energy Fuel 22(4):2775–2781
49. Hsu TC, Guo GL, Chen WH, Hwang WS (2010) Effect of dilute acid pretreatment of rice straw on structural properties and enzymatic hydrolysis. Bioresour Technol 101:4907–4913
50. Humpert D, Ebrahimi M, Czermak P (2016) Membrane technology for the recovery of lignin: a review. Membranes 6(3):42
51. Hussin MH, Rahim AA, Ibrahim MN, Perrin D, Brosse N (2015) Enhanced properties of oil palm fronds (OPF) lignin fractions produced via tangential ultrafiltration technique. Ind Crops Prod 66:1–10
52. Imman S, Arnthong J, Burapatana V, Champreda V, Laosiripojana N (2014) Effects of acid and alkali promoters on compressed liquid hot water pretreatment of rice straw. Bioresour Technol 171:29–36
53. Isikgor FH, Becer CR (2015) Lignocellulosic biomass: a sustainable platform for the production of bio-based chemicals and polymers. Polym Chem 6:4497–4559
54. Islam MS, Kao N, Bhattacharya SN, Gupta R, Bhattacharjee PK (2017) Effect of low pressure alkaline delignification process on the production of nanocrystalline cellulose from rice husk. J Taiwan Inst Chem Eng 80:820–834
55. Jablonský M, Haz A (2016) Characterization of lignins precipitated with nitric and hydrochloric acid. Int J Recent Sci Res 7:9177–9180
56. Jahan MS, Rahman MM, Sutradhar S, Quaiyyum MA (2015) Fractionation of rice straw for producing dissolving pulp in biorefinery concept. Nord Pulp Pap Res J 30(4):562–567

57. Jönsson AS, Nordin AK, Wallberg O (2008) Concentration and purification of lignin in hardwood kraft pulping liquor by ultrafiltration and nanofiltration. Chem Eng Res Des 86(11):1271–1280
58. Karimi K, Emtiazi,G, Taherzadeh MJ (2006) Ethanol production from dilute-acid pretreated rice straw by simultaneous saccharification and fermentation with *Mucor indicus*, *Rhizopus oryzae*, and *Saccharomyces cerevisiae*. Enzyme Microb Technol 40(1):138–144
59. Kaur A, Kuhad RC (2019) Valorization of rice straw for ethanol production and lignin recovery using combined acid-alkali pre-treatment. Bioenergy Res 12:570–582
60. Khaleghian H, Karimi K, Behzad T (2015) Ethanol production from rice straw by sodium carbonate pretreatment and Mucor hiemalis fermentation. Ind Crops Prod 76:1079–1085
61. Klett AS, Chappell PV, Thies MC (2015) Recovering ultraclean lignins of controlled molecular weight from kraft black-liquor lignins. Chem Commun 51(64):12855–12858
62. Knauf M, Moniruzzaman M (2004) Lignocellulosic biomass processing: a perspective. Sugar Tech 106(1263):147–150
63. Kordowska-Wiater M, Targoński Z (2002) Ethanol fermentation on glucose/xylose mixture by co-cultivation of restricted glucose catabolite repressed mutants of *Pichia stipitis* with respiratory deficient mutants of *Saccharomyces cerevisiae*. Acta Microbiol Pol 51(4):345–352
64. Kun N, Peng C, Xu Z, Wen-Song T (2009) Enhanced enzymatic hydrolysis of rice straw pretreated by alkali assisted with photocatalysis technology. J Chem Technol Biotechnol 84:1240–1245
65. Laplace, JM, Delgenes JP, Moletta R, Navarro JM (1993) Ethanol production from glucose and xylose by separated and co-culture processes using high cell density systems. Process Biochem 28:519–525
66. Lau MW, Gunawan C, Balan V, Dale BE (2010) Comparing the fermentation performance of *Escherichia coli* KO11, *Saccharomyces cerevisiae* 424A (LNH-ST) and Zymomonas mobilis AX101 for cellulosic ethanol production. Biotechnol Biofuels 3(1):1–10
67. Li H, Cai X, Wang Z, Xu C (2020) Cost-effective production of organosolv lignin from woody biomass using ethanol-water mixed solvent at mild conditions. J Supercrit Fluids 158:104745
68. Li MF, Sun SN, Xu F, Sun RC (2012) Sequential solvent fractionation of heterogeneous bamboo organosolv lignin for value-added application. Sep Purif Technol 101:18–25
69. Liao JJ, Abd Latif NH, Trache D, Brosse N, Hussin MH (2020) Current advancement on the isolation, characterization and application of lignin. Int J Biol Macromol 62:985–1024
70. Lim WL, Gunny AA, Kasim FH, AlNashef IM, Arbain D (2019) Alkaline deep eutectic solvent: a novel green solvent for lignocellulose pulping. Cellulose 26:4085–4098
71. Lo CC, Chang YW, Chen YL, Liu YL, Wu HS, Sun YM (2021) Lignin recovery from rice straw biorefinery solid waste by soda process with ethylene glycol as co-solvent. J Taiwan Inst Chem Eng 126:50–57
72. Lourenço A, Pereira H (2018) Compositional variability of lignin in biomass. In: Lignin-trends and applications, pp 65–98
73. Lourençon TV, Hansel FA, Da Silva TA, Ramos LP, De Muniz GI, Magalhães WL (2015) Hardwood and softwood kraft lignins fractionation by simple sequential acid precipitation. Sep Purif Technol 154:82–88
74. Lu Y, Lu YC, Hu HQ, Xie FJ, Wei XY, Fan X (2017) Structural characterization of lignin and its degradation products with spectroscopic methods. J Spectrosc 2017:1–5
75. Manorma, Ferreira I, Alves P, Gil MH, Gando-Ferreira LM (2021) Lignin separation from black liquor by mixed matrix polysulfone nanofiltration membrane filled with multiwalled carbon nanotubes. Sep Purif Technol 260:118231. https://doi.org/10.1016/j.seppur.2020.118231
76. Mateo S, Roberto IC, Sánchez S, Moya AJ (2013) Detoxification of hemicellulosic hydrolyzate from olive tree pruning residue. Ind Crops Prod 49:196–203
77. Maurya DP, Singla A, Negi S (2015) An overview of key pretreatment processes for biological conversion of lignocellulosic biomass to bioethanol. 3 Biotech 5(5):597–609
78. McIntosh S, Vancov T (2010) Enhanced enzyme saccharification of Sorghum bicolor straw using dilute alkali pretreatment. Bioresour Technol 101(17):6718–6727

79. Meng X, Parikh A, Seemala B, Kumar R, Pu Y, Wyman CE, Cai CM, Ragauskas AJ (2019) Characterization of fractional cuts of co-solvent enhanced lignocellulosic fractionation lignin isolated by sequential precipitation. Bioresour Technol 272:202–208
80. Meng X, Wells T, Sun Q, Huang F, Ragauskas A (2015) Insights into the effect of dilute acid, hot water or alkaline pretreatment on the cellulose accessible surface area and the overall porosity of populus. Green Chem 17(8):4239–4246
81. Mohammadi-Rovshandeh J, Talebizadeh A, Rezayati-Charani P (2005) Pulping of rice straw by high boiling solvents in atmospheric pressure. Iran Polym J14:223–227
82. Mohapatra S, Pattathil S, Thatoi H (2017) Structural and functional characterization of two *Pennisetum sp.* biomass during ultrasono-assisted alkali pretreatment and enzymatic hydrolysis for understanding the mechanism of targeted delignification and enhanced saccharification. ACS Sustain Chem Eng 5(8):6486–6497
83. Moniz P, Lino J, Duarte LC, Roseiro LB, Boeriu CG, Pereira H, Carvalheiro F (2015) Fractionation of hemicelluloses and lignin from rice straw by combining autohydrolysis and optimised mild organosolv delignification. BioResources 10:2626–2641
84. Mosier N, Wyman C, Dale B, Elander R, Lee YY, Holtzapple M, Ladisch M (2005) Features of promising technologies for pretreatment of lignocellulosic biomass. Bioresour Technol 96(6):673–686
85. Nguyen TAD, Kim KR, Han SJ, Cho HY, Kim JW, Park SM, Park JC, Sim SJ (2010) Pretreatment of rice straw with ammonia and ionic liquid for lignocellulose conversion to fermentable sugars. Bioresour Technol 101(19):7432–7438
86. Novo LP, Curvelo AA (2019) Hansen solubility parameters: a tool for solvent selection for organosolv delignification. Ind Eng Chem Res 58(31):14520–14527
87. Nowakowski DJ, Jones JM (2008) Uncatalysed and potassium-catalysed pyrolysis of the cell-wall constituents of biomass and their model compounds. J Anal Appl Pyrolysis 83:12–25
88. Oberoi HS, Babbar N, Sandhu SK, Dhaliwal SS, Kaur U, Chadha BS, Bhargav VK (2012) Ethanol production from alkali-treated rice straw via simultaneous saccharification and fermentation using newly isolated thermotolerant *Pichia kudriavzevii* HOP-1. J Ind Microbiol Biotechnol 39(4):557–566
89. Oberoi HS, Vadlani PV, Brijwani K, Bhargav VK, Patil RT (2010) Enhanced ethanol production via fermentation of rice straw with hydrolysate-adapted *Candida tropicalis* ATCC 13803. Process Biochem 45(8):1299–1306
90. Olivares MA, Guzmán JA, Natho AR, Saavedra A (1988) Kraft lignin utilization in adhesives. Wood Sci Technol 22(2):157–165
91. Pan X, Gilkes N, Kadla J, Pye K, Saka S, Gregg D, Ehara K, Xie D, Lam D, Saddler J (2006) Bioconversion of hybrid poplar to ethanol and co-products using an organosolv fractionation process: optimization of process yields. Biotechnol Bioeng 94(5):851–861
92. Pan XJ, Sano Y, Ito T (1999) Atmospheric acetic acid pulping of rice straw II: behavior of ash and silica in rice straw during atmospheric acetic acid pulping and bleaching. Holzforschung 53(1):49–55
93. Panagiotopoulos IA, Bakker RR, Vrije TD, Claassen PAM, Koukios EG (2012) Integration of first and second generation biofuels: fermentative hydrogen production from wheat grain and straw. Bioresour Technol 128:345–350
94. Pang T, Wang G, Sun H, Sui W, Si C (2021) Lignin fractionation: effective strategy to reduce molecule weight dependent heterogeneity for upgraded lignin valorization. Ind Crops Prod 165:113442
95. Park Y, Doherty WOS, Halley PJ (2008) Developing lignin based resin coatings and composites. Ind Crop Prod 27(2):163–167
96. Park JY, Shiroma R, Al-Haq MI, Zhang Y, Ike M, Arai-Sanoh Y, Ida A, Kondo M, Tokuyasu K (2010) A novel lime pretreatment for subsequent bioethanol production from rice straw–calcium capturing by carbonation (CaCCO) process. Bioresour Technol 101(17):6805–6811
97. Patel H, Divecha J, Shah A (2017) Microwave assisted alkali treated wheat straw as a substrate for co-production of (hemi)cellulolytic enzymes and development of balanced enzyme cocktail for its enhanced saccharification. J Taiwan Inst Chem Eng 71:298–306

98. Ponnuchamy V, Gordobil O, Diaz RH, Sandak A, Sandak J (2021) Fractionation of lignin using organic solvents: a combined experimental and theoretical study. Int J Biol Macromol 168:792–805
99. Pradyawong S, Shrestha R, Li P, Sun XS, Wang D (2021) Effect of pH and pH-shifting on lignin–protein interaction and properties of lignin-protein polymers. J Polym Environ 1:1–12
100. Priyanto S, Intan AP, Rianto B, Kusworo TD, Pramudono B, Untoro E, Ratu P (2019) The effect of acid concentration (H_2SO_4) on the yield and functional group during lignin isolation of biomass waste pulp and paper industry. Reaktor 19(4):162–167
101. Pu Y, Ziemer C, Ragauskas AJ (2006) CP/MAS 13C NMR analysis of cellulase treated bleached softwood kraft pulp. Carbohydr Res 341(5):591–597
102. Punnapayak H, Emert GH (1986) Use of *Pachysolen tannophilus* in simultaneous saccharification and fermentation (SSF) of lignocellulosics. Biotechnol Lett 8:63–66
103. Ragauskas AJ, Beckham GT, Biddy MJ, Chandra R, Chen F, Davis MF, Davison BH, Dixon RA, Gilna P, Keller M, Langan P (2014) Lignin valorization: improving lignin processing in the biorefinery. Science 344(6185):1246843
104. Rahnama N, Mamat S, Shah UK, Ling FH, Rahman NA, Ariff AB (2013) Effect of alkali pretreatment of rice straw on cellulase and xylanase production by local *trichoderma harzianum* SNRS3 under solid state fermentation. BioResources 8(2):2881–2896
105. Rastogi M, Shrivastava S (2017) Recent advances in second generation bioethanol production: an insight to pretreatment, saccharification and fermentation processes. Renew Sustain Energy Rev 80:330–340
106. Roberto IC, Mussatto SI, Rodrigues RC (2003) Dilute-acid hydrolysis for optimization of xylose recovery from rice straw in a semi-pilot reactor. Ind Crops Prod 7:171–176
107. Rodrigues JS, Lima V, Araujo LC, Botaro VR (2021) Lignin fractionation methods: can lignin fractions be separated in a true industrial process? Ind Eng Chem Res 60(30):10863–10881
108. Rodriguez A, Serrano L, Moral A, Jimenez L (2008) Pulping of rice straw with high-boiling point organosolv solvents. Biochem Eng J 42:243–247
109. Rodriguez A, Moral A, Serrano L, Labidi J, Jimenez L (2008) Rice straw pulp obtained by using various methods. Bioresour Technol 99:2881–2886
110. Sadeghifar H, Ragauskas A (2020) Perspective on technical lignin fractionation. ACS Sustain Chem Eng 8(22):8086–8101
111. Saito T, Perkins JH, Vautard F, Meyer HM, Messman JM, Tolnai B, Naskar AK (2014) Methanol fractionation of softwood kraft lignin: Impact on the lignin properties. ChemSusChem 7(1):221–228
112. Sambusiti C, Ficara E, Malpei F, Steyer JP, Carrere H (2013) Effect of sodium hydroxide pretreatment on physical, chemical characteristics and methane production of five varieties of sorghum. Energy 55:449–456
113. Sarartale GD, Saratale RG, Ghodake GS, Jiang YY, Chang JS, Shin HS, Kumar G (2017) Solid state fermentative lignocellulolytic enzymes production, characterization and its application in the saccharification of rice waste biomass for ethanol production: an integrated biotechnological approach. J Taiwan Inst Chem Eng 76:51–58
114. Schutyser W, Renders AT, Van den Bosch S, Koelewijn SF, Beckham GT, Sels BF (2018) Chemicals from lignin: an interplay of lignocellulose fractionation, depolymerisation, and upgrading. Chem Soc Rev 47(3):852–908
115. Silva JPA, Mussatto SI, Roberto IC, Teixeira JA (2011) Ethanol production from xylose by *Pichia stipitis* NRRL Y-7124 in a stirred tank bioreactor. Braz J Chem Eng 28(1):151–156
116. Singh G, Arya SK (2020) A review on management of rice straw by use of cleaner technologies: abundant opportunities and expectations for Indian farming. J Clean Prod 291:125278
117. Singh R, Srivastava M, Shukla A (2015) Environmental sustainability of bioethanol production from rice straw in India: a review. Renew Sust Energ Rev 54:202–216
118. Singh YD, Satapathy KB (2018) Conversion of lignocellulosic biomass to bioethanol: an overview with a focus on pretreatment. Int J Eng Technol 15:17–43

119. Soudham VP, Brandberg T, Mikkola JP, Larsson C (2014) Detoxification of acid pretreated spruce hydrolysates with ferrous sulfate and hydrogen peroxide improves enzymatic hydrolysis and fermentation. Bioresour Technol 166:559–565
120. Spindler DD, Wyman CE, Mohagheghi A, Grohmann K (1988) Thermotolerant yeast for simultaneous saccharification and fermentation of cellulose to ethanol. Appl Biochem Biotechnol 17:279–293
121. Sreekumar A, Shastri Y, Wadekar P, Patil M, Lali A (2020) Life cycle assessment of ethanol production in a rice-straw-based biorefinery in India. Clean Technol Environ Policy 22:409–422
122. Stewart H, Golding M, Merino LM, Archer R, Davies C (2014) Manufacture of lignin microparticles by anti-solvent precipitation: effect of preparation temperature and presence of sodium dodecyl sulfate. Food Res Int 66:93–99
123. Su CH, Chung MH, Hsieh HJ, Chang YK, Ding JC, Wu HM (2012) Enzymatic hydrolysis of lignocellulosic biomass in ionic liquid media for fermentable sugar production. J Taiwan Inst Chem Eng 43:573–577
124. Sulbarán-de-Ferrer B, Aristiguieta M, Dale BE, Ferrer A, Ojeda-de-Rodriguez G (2003) Enzymatic hydrolysis of ammonia-treated rice straw. Appl Biochem Biotechnol 105:155–164
125. Sumphanwanich J, Leepipatpiboon N, Srinorakutara T, Akaracharanya A (2008) Evaluation of dilute-acid pretreated bagasse, corn cob and rice straw for ethanol fermentation by *saccharomyces cerevisiae*. Ann Microbiol 58:219–225
126. Sun RC, Tomkinson J, Ma PL, Liang SF (2000) Comparative study of hemicelluloses from rice straw by alkali and hydrogen peroxide treatments. Carbohyd Polym 42(2):111–122
127. Sun S, Chen W, Tang J, Wang B, Cao X, Sun S (2016) Synergetic effect of dilute acid and alkali treatments on fractional application of rice straw. Biotechnol Biofuel 9(1):1–13
128. Šurina I, Jablonský M, Ház A, Sladková A, Briškárová A, Kačík F, Šima J (2015) Characterization of non-wood lignin precipitated with sulphuric acid of various concentrations. BioRes 10(1):1408–1423
129. Taherzadeh MJ, Niklasson C (2004) Ethanol from lignocellulosic materials: pretreatment, acid and enzymatic hydrolyses, and fermentation. In: ACS symposium series, vol 889, pp 49–68
130. Tan IS, Lee KT (2014) Enzymatic hydrolysis and fermentation of seaweed solid wastes for bioethanol production: an optimization study. Energy 78:53–62
131. Taniguchi M, Itaya T, Tohma T, Fujii M (1997) Ethanol production from a mixture of glucose and xylose by a novel co-culture system with two fermentors and two microfiltration modules. J Ferment Bioeng 84(1):59–64
132. Tarkov H, Feist WC (1969) A mechanism for improving the digestibility of lignocellulosic materials with dilute alkali and liquid ammonia. Adv Chem Ser 95:197–218
133. Teramura H, Sasaki K, Oshima T, Aikawa S, Matsuda F, Okamoto M (2015) Changes in lignin and polysaccharide components in 13 cultivars of rice straw following dilute acid pretreatment as studied by solution-state 2D ^{1}H–^{13}C NMR. PLoS ONE 10(6):e0128417
134. Toledano A, Serrano L, Garcia A, Mondragon I, Labidi J (2010) Comparative study of lignin fractionation by ultrafiltration and selective precipitation. Chem Eng J 157(1):93–99
135. Toledano A, Serrano L, Labidi J (2012) Process of olive tree pruning lignin revalorization. Chem Eng J 193(3):396–403
136. Tortora M, Cavalieri F, Mosesso P, Ciaffardini F, Melone F, Crestini C (2014) Ultrasound driven assembly of lignin into microparticles for storage and delivery of hydrophobic molecules. Biomacromolecules 15:1634–1643
137. Valdes G, Planes LR (1983) Study of the hydrolysis of rice straw with sulfuric acid under moderate conditions. Rev Cienc Quim 14:11–19
138. Valkanas GN, Valkanas NP, Vlyssides AG, Theodoropoulos AG (1998) Method for production of ethyl alcohol. USP 5766895
139. Velazquez-Lucio J, Rodríguez-Jasso RM, Colla LM, Sáenz-Galindo A, Cervantes-Cisneros DE, Aguilar CN, Fernandes BD, Ruiz HA (2018) Microalgal biomass pretreatment for bioethanol production: a review. Biofuel Res J 5(1):780–791

140. Wang G, Chen H (2013) Fractionation of alkali-extracted lignin from steam-exploded stalk by gradient acid precipitation. Sep Purif Technol 105:98–105
141. Wang M, Zhou D, Wang Y, Wei S, Yang W, Kuang M, Ma L, Fang D, Xu S, Du S (2016) Bioethanol production from cotton stalk: a comparative study of various pretreatments. Fuel 184:527–532. https://doi.org/10.1016/j.fuel.2016.07.061
142. Wang T, Jiang M, Yu X, Niu N, Chen L (2022) Application of lignin adsorbent in wastewater treatment: a review. Sep Purif Technol 16:122116
143. Williams T, Hosur M, Theodore M, Netravali A, Rangari V, Jeelani S (2011) Time effects on morphology and bonding ability in mercerized natural fibers for composite reinforcement Int J Polym Sci 1–9
144. Wiselogel A, Tyson S, Johnson D (1996) Biomass feedstock resources and composition. In: Wyman CE (ed) Handbook on bioethanol: production and utilization (applied energy technology series), pp 105–118
145. Wu J, Dong L, Liu B, Xing D, Zhou C, Wang Q, Wu X, Feng L, Cao G (2020) A novel integrated process to convert cellulose and hemicellulose in rice straw to biobutanol. Environ Res 186:109580
146. Wunna K, Nakasaki K, Auresenia JL, Abella LC, Gaspillo PD (2017) Effect of alkali pretreatment on removal of lignin from sugarcane bagasse. Chem Eng Trans 56:1831–1836
147. Wyman CE (1994) Ethanol from lignocellulosic biomass: technology, economics, and opportunities. Bioresour Technol 50(1):3–15
148. Xin D, Yang Z, Liu F, Xu X, Zhang J (2015) Comparison of aqueous ammonia and dilute acid pretreatment of bamboo fractions: structure properties and enzymatic hydrolysis. Bioresour Technol 175:529–536
149. Xin LZ, Kumakura M (1993) Effect of radiation pretreatment on enzymatic hydrolysis of rice straw with low concentrations of alkali solution. Bioresour Technol 43(1):13–17
150. Xue BL, Wen JL, Sun RC (2014) Lignin based rigid polyurethane foam reinforced with pulp fiber: synthesis and characterization. ACS Sustain Cheml Eng 1:1474–1480
151. Yang H, Shi Z, Xu G, Qin Y, Deng J, Yang J (2019) Bioethanol production from bamboo with alkali-catalyzed liquid hot water pretreatment. Bioresour Technol 274:261–266
152. Yang L, Cao J, Jin Y, Chang H-M, Jameel H, Phillips R, Li Z (2012) Effects of sodium carbonate pretreatment on the chemical compositions and enzymatic saccharification of rice straw. Bioresour Technol 124:283–291
153. Yin F, Hwang A, Yu N, Hao P (1982) Hydrolysis of agricultural wastes. In: Proceedings of resource recovery solid wastes conference, pp 447–456
154. Yucel HG, Aksu Z (2015) Ethanol fermentation characteristics of *Pichiastipitis* yeast from sugar beet pulp hydrolysate: use of new detoxification methods. Fuel 158:793–799
155. Zabed H, Sahu JN, Boyce AN, Faruq G (2016) Fuel ethanol production from lignocellulosic biomass: an overview on feedstocks and technological approaches. Renew Sust Energ Rev 66:751–774
156. Zhang QZ, Cai WM (2008) Enzymatic hydrolysis of alkali-pretreated rice straw by *Trichoderma reesei* ZM4-F3. Biomass Bioenergy 32:1130–1135
157. Zhong C, Lau MW, Balan V, Dale E, Yuan YJ (2009) Optimization of enzymatic hydrolysis and ethanol fermentation from AFEX-treated rice straw. Appl Microbiol Biotechnol 84:667–676
158. Zhou N, Thilakarathna WW, He QS, Rupasinghe HV (2022) A review: depolymerization of lignin to generate high-value bio-products: opportunities, challenges, and prospects. Front Energy Res 9:758744
159. Zhu S, Wu Y, Zhao Y, Tu S, Xue Y (2006) Fed-batch simultaneous saccharification and fermentation of microwave/acid/alkali/H_2O_2 pretreated rice straw for production of ethanol. Chem Eng Commun 193:639–648
160. Zhu S, Wu Y, Yu Z, Liao J, Zhang Y (2005) Pretreatment by microwave/alkali of rice straw and its enzymic hydrolysis. Process Biochem 40(9):3082–3086

Facts on Climatic Effect of Waste Paper Recycling in the Pulp and Paper Industry

Saakshy and Rahul Mishra

Abstract The recycling of pulp and paper is said to be a key solution of mitigating environmental impact, carbon emissions and promoting sustainability. The chapter makes a review of studies conducted on aspect of recycling of paper which is said to decrease the cutting of trees and deforestation and increase ecological aspects. It has been observed in the previous studies that environment load decreases on the utilization of recycled paper. However, some studies also pointed that recycling of paper could also increase green house gas emissions if fossil fuels are used in place of renewable energy. The high energy demands of the recycling process, varied grade of recycled grades of paper and dependency on virgin raw material for specific grades of paper make the limited utilization of recyled fibre. Moreover, the energy intensive recycling process sometimes increase the carbon footprint. The recycling of pulp and paper remains a topic of study for circular economy in the pulp and paper industry. The actual climatic benefits of global recycling of pulp and paper remains limited. The holistic approach of utilization of renewable energy, improved forest policies along with recycling of paper can be an option to radically reduce the green house gas emissions in the pulp and paper sector.

Keywords Pulp and paper industry · Global recycling · Climatic benefits · Carbon dioxide emission · Waste reduction · Environmental impact · Greenhouse gas inventories

1 Introduction

The Indian paper industry produces about 5% of the worlds total paper production. The domestic market size of Indian paper industry is about Rs. 80,000/- crore. During the last decade, from 2012–13 to 2022–2023, India's paper consumption registered a CAGR (Compound Annual Growth Rate) of 6% compared to the global growth of

Saakshy (✉) · R. Mishra
Kumarappa National Handmade Paper Institute, Jaipur, India
e-mail: saakshyagarwal@gmail.com

S. Sundaramurthy et al. (eds.), *Black Liquor Waste Management in the Pulp and Paper Industry*, Sustainable Materials and Technology,
https://doi.org/10.1007/978-981-96-4237-3_9

3% making India one of the fastest growing paper producers in the world [Press Trust of India 7th May 2022]. The Indian paper industry is highly dependent on the raw materials such as wood, bamboo, and bagasse. The shortage or price fluctuation of raw material adversely affect the sustainability of the paper industry including increase in production costs, logistics, chemicals, and raw materials. Due to the pandemic and Ukraine war, the entire paper business community turned mindful of the supply chain scenario for 2023. The coal prices have also been impacted by Indonesia's decision to curb exports. The issue of a conspicuous shortage of basic raw materials together with their high cost has aggravated the pressure on the margins and the viability of the operations. The gravity of the situation is reflected in the fact that, at present the cost of procuring wood in India is more than $100 per tonne whereas the same in other competing countries that have advanced paper manufacturing is $60 per tonne. India's share in the demand for paper across the globe has seen impressive growth. Growing manufacturing sector, requirement of better quality packaging of FMCG products and the demand for the upstream market of paper products, such as tissue paper, filter paper, tea bags, light weight online coated paper and medical grade coated paper are expected to drive the paper & paper products market in India in coming years. The per capita paper consumption in India is at a little over 13 kg, is way behind the global average of 57 kg. Under the baseline scenario, by 2024–25, domestic consumption is projected to rise to 23.5-million tpa and production to 22.0 million tonnes per annum. About one million tpa of integrated pulp, paper and paperboard capacity is required to be created in India on an annual basis over the current capacity to meet the growing demand (Paprex 2023). According to CPPRI data (2019–2020), there are approximately 526 operational mills in India. Globally, there is emphasis on increase in recycling of paper and paper waste as a raw material for making paper and paper products. The present market of Indian paper and paper products is expected to reach $13.4 billion by 2024.

The concept of a "Recycling" has gained significant attention and momentum in recent years. Paper recycling has a positive impact on the environment and can help mitigate climate change. According to a study published in IntechOpen, paper recycling conserves natural resources, decreases the operation and capital costs of paper production, and reduces the environmental pollution. However, it is important to note that the recycling process can also have negative environmental impacts if it relies on non-renewable energy resources such as fossil fuels. Therefore, it is crucial to use renewable energy sources to power the recycling process and achieve the sustainable development goals.

There has been a decline in the export of recycled paper from India and China, as mills have come up in Turkey, EU, Indonesia, Vietnam, and Malaysia [30]. The Bureau of International Recycling (BIR) reports that there is a lack of demand for recovered paper across all parts of the world, which has led to a decline in demand for packaging material [3]. India is a world leader in waste paper recycling, and the Indian recycling industry needs nearly 14 million tonnes of waste paper to meet the current demands [19]. The United States is the largest exporter of waste paper to India, while the United Kingdom and the rest of Europe come in second with 17 and 18%, respectively [21].

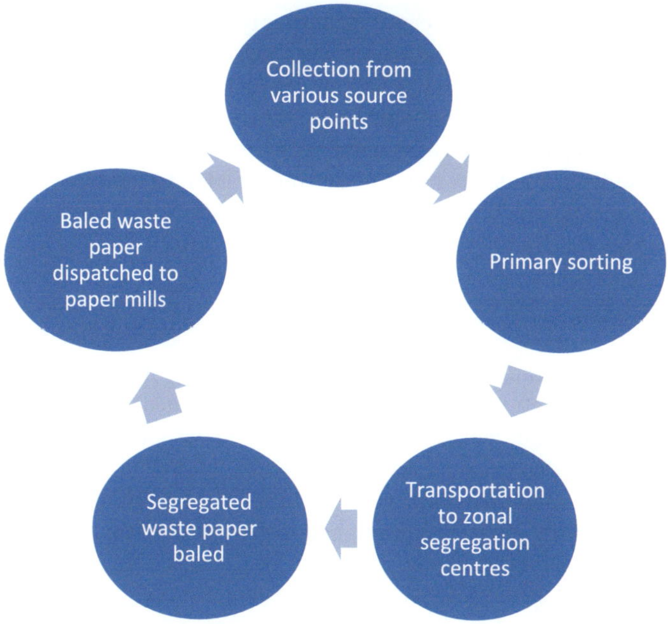

Fig. 1 Flow diagram of the paper recycling process

The material flow analysis in combination with input–output analysis study measured the waste paper recovery rate and analyzed the impact of waste paper recycling on the carbon emissions from China's paper industry. China's waste paper recovery rate estimated in this study was close to 70% in 2017, much higher than that of 48% obtained with the traditional method. Thus, although China's waste paper recovery rate has reached a relatively high level, currently, waste paper recycling is still an effective method to reduce carbon emissions from China's paper industry [25].

The collections & recovery process for waste management begins with direct collectors from various source points and shops where primary sorting of the waste into different materials occurs. Then, the waste is transported to zonal segregation centres, where the material is collected from shops and baled for dispatch to the end users. The Indian waste paper collection is mostly carried out by the informal sector, rag pickers and door to door collectors (Fig. 1).

2 Circular Economy

Circular economy is a model of production and consumption that involves sharing, leasing, reusing, repairing, refurbishing, and recycling of the existing materials. The circular economy aims to eliminate waste, keep products and materials in use, and regenerate natural systems. It is based increasingly on renewable energy and materials and is accelerated by digital innovation. The circular economy is a resilient, distributed, diverse, and inclusive economic model that tackles climate change and other global challenges, such as biodiversity loss, handling of waste and pollution by decoupling economic activity from the consumption of finite resources (Fig. 2).

The pulp and paper industry has been actively embracing circular economy practices to maximize the use of resources and reduce waste. Some examples of circular economy practices in the pulp and paper industry include utilization of waste streams, maximization of resources, valorization of industrial waste and transformation of paper industry.

The industry is negating the exploitation of virgin resources and raw materials by utilizing the waste streams. This practice reduced waste generation and contributed to the circularity of the industry. The pulp and paper industry aims to maximize the use of resources and reduce waste by making the best possible use of products. This includes efforts to extend the life of products, promote reuse, refurbishment, and recycling, and minimize resource extraction [20]. The industry is focusing on getting value from the pulp and paper industry wastes, aiming to move towards sustainability and circular economy practices. This involves the reuse and recycling of resources instead of the dominant linear economic strategy of "take, make, and dispose" [2]. The pulp and paper industry is transforming its operations to align with circular economy principles. This includes reducing waste generation, increasing the use of by-products to replace virgin materials, and seeking innovative ways to improve resource efficiency [5].

These examples demonstrate the industry's commitment to adopting circular economy practices, such as waste reduction, resource optimization, and the valorization of industry wastes, to create a more sustainable and regenerative economic system.

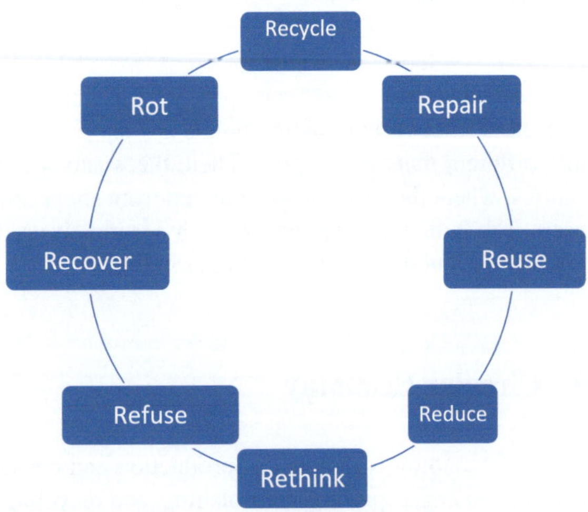

Fig. 2 8 R's of circular economy

3 Principles of Circular Economy

The key principles of a circular economy include the elimination of waste and pollution, circulation of products and materials and regeneration of nature. The circular economy aims to design out waste and pollution by rethinking the supply chain and challenging businesses to avoid waste and pollution. This principle focuses on reducing the environmental impact of production and consumption processes. This principle emphasizes keeping products and materials in use at their highest value for as long as possible. It involves processes such as maintenance, reuse, refurbishment, remanufacture, recycling, and composting. By extending the life of products and materials, the circular economy aims to minimize resource extraction and waste generation. The circular economy seeks to regenerate natural systems by promoting sustainable production and consumption of goods. It advocates for the preservation or enhancement of renewable resources, such as returning valuable nutrients to the soil to support regeneration and using renewable energy instead of relying on fossil fuels.

These principles underpin the transition to a circular economy, which aims to decouple economic activity from the consumption of finite resources and create a more sustainable and regenerative economic system. The circular economy model is designed to address global challenges such as climate change, biodiversity loss, waste, and pollution by promoting a more efficient and environmentally friendly approach to production and consumption [8, 34]. The 10 principles of circular economy include refuse, rethink, reduce, reuse, repair, refurbish, remanufacture, repurpose, recycle and recover [24]. The principles of circular economy also include preservation and optimization of natural resources, reducing consumption of natural resources, achieving eco-design, development of local economy, putting an end to obsolescence, reducing impact of human activity on environment and health and limiting consumption.

4 Environmental Challenges

The paper industry is a significant contributor to environmental pollution, and therefore, it is subject to stringent environmental regulations. Many environmental issues arise from the operation of a pulp and paper mill, such as climate change, human toxicity, eco-toxicity, photochemical oxidation, acidification, nutrification, deforestation and solid refuse production. The Green Rating Project (GRP), undertaken by the Centre for Science and Environment, New Delhi attempts to examine the environmental impact in terms of raw material inputs, fiber sourcing, discharges and uses to formulate a holistic and broad-based criterion of environmental parameters factoring all these aspects. Globally, the paper and pulp business were responsible for approximately 190 million metric tonnes (Mt) of CO_2 emissions in 2021, which was a record high for the sector and accounted for approximately 2% of all industrial emissions.

According to the estimates published in the Global Carbon Budget Report in 2022, India had the highest rate of carbon emission growth and is the main contributor to global warming. It increased by 3.8% annually to become the fourth largest emitter worldwide. India's per capita carbon emissions are still one-eighth of those of the United States and one-third of the global average. According to the Report, India's total emissions in 2021 are estimated to be 2.7 billion tons, marginally less than the EU's total of 2.8 billion tons. India's share of global emissions in 2021 was 7.5%, marginally lower than the EU's 7.7% [29].

5 Main Sources of Carbon Dioxide Emissions in India

The main sources of carbon dioxide (CO_2) emissions in India include the energy sector, agriculture, industrial processes etc. In India, 68.7% of greenhouse gas (GHG) emissions come from the energy sector, which is primarily driven by the combustion of fossil fuels for energy production. Coal combustion in India produced approximately 1.8 billion metric tons of CO_2 in 2021, representing an increase of roughly 13% from the previous year. Agriculture is another significant source of CO_2 emissions in India, with rice paddies being a major contributor. Methane emissions from livestock and rice production, as well as leakages from oil and gas production, also contribute to India's greenhouse gas emissions. Industrial processes, including manufacturing and construction, contribute to India's CO_2 emissions. The production and use of iron and steel, for example, are significant sources of CO_2 emissions [31]. Land-use change, including deforestation and forest degradation, also contribute to India's greenhouse gas emissions. These sources of CO_2 emissions highlight the challenges associated with reducing emissions and transitioning to more sustainable energy practices in India.

The average carbon footprint per person in the world varies across different sources. According to the Nature Conservancy, the global average carbon footprint is closer to 4 tons per person [27]. This figure is supported by the International Energy Agency (IEA), which states that the global average energy-related carbon footprint is around 4.7 tonnes of CO_2 per person [3]. World in Data reports that the global average per capita CO_2 emissions were 4.8 tonnes per person in 2017 [16]. It's important to note that there are significant disparities in per capita emissions across different countries. For instance, the United States has one of the highest average carbon footprints, with an average of 16 tons per person, while India's average carbon footprint is approximately 1.89 tons per person [37]. These variations are influenced by factors such as income, lifestyle, consumption patterns, and energy use. In summary, the average carbon footprint per person in the world is estimated to be around 4 tons of CO_2 emissions [7, 12].

6 Paper Life Cycle System

The life cycle of paper involves several stages including deforestation, grinding, filtration, heating and recycling i.e. from the extraction of raw materials to the disposal of used paper. The first stage of the paper life cycle involves the cutting down of trees, which are then transported to a paper mill for processing. The wood obtained from these trees is used to make pulp. The wood is then chopped into small pieces and transformed into wood pulp with the help of chemicals and water. The pulp is then filtered to remove the impurities and other unwanted materials. The pulp is then heated and pressed to remove the excess water and compress the pulp into paper sheets. Once the paper has been used for its intended purpose, it can be recycled. The used paper is collected, sorted, and cleaned before being processed into new paper products. Recycled paper can go through the recycling process up to 6–7 times before the cellulose fibers become incapable of forming sheets. The recycling of paper is an important part of the circular economy, as it helps to reduce waste and minimize the environmental impact of paper production [4, 9] (Fig. 3).

Fig. 3 Flow chart of the paper life cycle

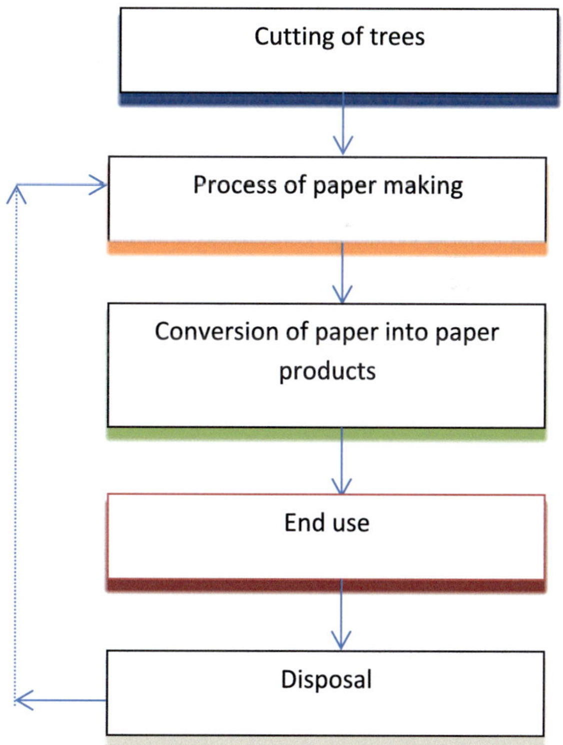

7 Environmental Impacts of Paper Production

The environmental impacts of paper production are significant and include various environmental issues. Some of the main impacts include deforestation, air and water pollution, green house gas emissions and solid waste. Paper production relies heavily on the use of wood, which contributes to deforestation. Deforestation leads to habitat loss, species extinction, and climate change [36]. The manufacturing process of paper releases nitrogen oxide, sulfur oxide, and carbon dioxide into the air, contributing to pollution such as acid rain [26]. Additionally, paper mills can discharge various pollutants into surrounding water bodies, causing damage to aquatic ecosystems and threatening the health of people living near the mills. The production, transportation, and disposal of paper contribute to greenhouse gas emissions. When paper rots, it emits methane, a potent greenhouse gas, while combustion or composting releases carbon dioxide [6]. Discardarded paper and paperboard make up roughly 26% of solid municipal waste in landfill sites. There is a growing trend towards sustainability in the pulp and paper industry, aiming to reduce clear cutting, water use, greenhouse gas emissions, fossil fuel consumption, and pollution in local water supplies and air.

8 Carbon Dioxide Emissions from Fuel Combustion in India

India's carbon dioxide (CO_2) emissions from fuel combustion have been a significant concern due to their impact on climate change. According to data from Statista, coal combustion in India produced approximately 1.8 billion metric tons of carbon dioxide ($GtCO_2$) in 2021, representing an increase of roughly 13% from the previous year [31]. These emissions are primarily dominated by the burning of fossil fuels for energy production and industrial activities [31]. Enerdata's Global Energy and Climate Trends report indicates a new surge in CO_2 emissions in India, with an increase of 8.8% in coal-dependent emissions in 2022 [10]. The report also highlights that global energy-related CO_2 emissions increased by 2.5% in 2022, reaching a record high of over 33.8 $GtCO_2$, despite the global economic slowdown [3]. The World Bank's data on CO_2 emissions from transport in India shows the percentage of total fuel combustion attributed to transport-related emissions [35]. This data provides insights into the contribution of the transportation sector to India's overall CO_2 emissions. The International Energy Agency (IEA) also reports that global CO_2 emissions from energy combustion and industrial processes grew by 0.9% in 2022, reaching a new all-time high of 36.8 $GtCO_2$. The report highlights that, emissions growth in 2022 was below the global GDP growth, reverting to a decades-long trend of decoupling emissions and economic growth that was broken in 2021 [1]. These sources provide a comprehensive overview of India's CO_2 emissions from fuel combustion, highlighting the challenges associated with reducing emissions and transitioning to more sustainable energy practices.

9 Guidelines for National Green House Gas Inventories in India

India's national greenhouse gas (GHG) inventory is prepared in accordance with the Intergovernmental Panel on Climate Change (IPCC) Guidelines for National Greenhouse Gas Inventories. The IPCC Guidelines provide detailed instructions for the application of various methods for the estimation of GHG removals and emissions from sinks and sources across all sectors, and on reporting to the Conference of the Parties (COP) [18]. India's national inventory uses Tier 2 emission factors for cattle. The country has adopted the 2006 IPCC Guidelines for National Greenhouse Gas Inventories, which provide a technically sound methodological basis for preparing national GHG inventories [28]. The 2019 Refinement to the 2006 IPCC Guidelines for National Greenhouse Gas Inventories was adopted and accepted during the 49th Session of the IPCC in May 2019 to ensure the latest scientific and technical advancements. This refinement was prepared by the Task Force on National Greenhouse Gas Inventories (TFI) in accordance with the decision taken at the 44th Session of IPCC in Bangkok, Thailand, in October 2016 [14]. India's GHG inventory process involves various sectors, including waste, forestry, and energy sectors. The country has also adopted the IPCC's best practices for inventory estimation and has implemented different rules to control plastic waste, e-waste, construction and demolition waste, and metals recycling [28].

10 Environmental Effects of Recycling

The recycling of pulp and paper, an important part of a circular economy, may yield limited climate benefits. A study focusing on the global paper life cycle, which generates 1.3% of global greenhouse gas emissions found that emission pathways consistent with a 2 °C global warming target require strong reductions in the carbon intensity of electricity and heat generation. The additional recycling yields small or negative climate change mitigation benefits when it requires high-carbon grid electricity and displaces virgin pulping that is powered by low-carbon pulping by-products. The impact of greater recycling on greenhouse gas emissions is dependent on how the paper sector responds to a decrease in electricity and heat from black liquor recovery. The study suggests that the potential climate benefits of global recycling of pulp and paper are limited and that the energy implications of recycling should be carefully considered [32], Recycling 1 kg of paper can reduce carbon emissions by approximately 1 ton CO_2/t paper. Increased recycling rates and the use of recovered fibres in paper production lead to a decrease in life-cycle energy use and carbon dioxide emissions. The average life-cycle energy use of paper produced in The Netherlands, with a recycling rate of approximately 75%, is about 14 GJ/t, resulting in CO_2 savings of about 1 t CO_2/t paper if no recycled fibres are used. Plausible calculations suggest that countries like the US or Australia could achieve substantial greenhouse gas reductions through increased recycling, particularly of paper [15, 33]. Recycling and waste reduction in the paper industry can contribute to a reduction in industrial energy use and emissions, leading to further carbon emission reductions.

India has been actively formulating policies and promoting projects to drive the country towards a circular economy. The government has notified various rules, such as the Plastic Waste Management Rules, e-Waste Management Rules, Construction and Demolition Waste Management Rules, and Metals Recycling Policy, to support the transition to a circular economy. To expedite this transition, committees have been formed to prepare comprehensive action plans for transitioning from a linear to a circular economy in their respective focus areas. These committees are led by concerned line ministries and comprise officials from the Ministry of Environment, Forests and Climate Change (MoEFCC), NITI Aayog, domain experts, academics, and industry representatives [22]. India's circular economy strategy envisions that industrial and consumer wastes can replace virgin materials, and it aims to capture at least 60% of the electronic waste generated in cities. Circular economy practices in India include reverse logistics, beneficial reuse of wasted materials or energy, and business model innovation. A recent study suggested that India can unlock approximately half-a-trillion dollars of economic value by 2030 through the adoption of circular economy business models [11]. The transition to a circular economy in India is estimated to result in an additional US$ 4.5 trillion in global economic output by 2030. India's circular economy development route might generate an annual value of US$ 218 billion by 2030 and US$ 624 billion by 2050. The country is well-positioned to use digital technology to create innovative and cutting-edge circular businesses, and it has a competitive advantage over mature economies in adopting circular methods of production and building sustainable designs [17]. The government of India has been actively formulating policies and various incentivizing projects to drive the nation towards a sustainable circular economy. It has implemented different rules to control plastic waste, e-waste, construction and demolition waste, and metals recycling [13]. NITI Aayog, the policy think tank of the Government of India, has established a Circular Economy Cell (CE Cell) to work in the area of circular economy. The CE Cell has finalized 10 sectoral Circular Economy action plans for implementation by stakeholder ministries/departments [23].

11 Conclusions

It is common belief that recycling may yield greater benefits in industries while the climate change mitigation challenge is altogether different for industries with non-energy green house gas emissions. Recycling of paper helps to reduce the greenhouse gas emissions that can contribute to climate change as it is considered that it takes 70% less energy and water to recycle paper than to create new paper from trees. So, manufacturing from recovered paper cuts down on pollution. However, some studies also conclude that full benefits of paper recycling can only be realized if the process is powered by renewable energy. The studies also conclude that if all the waste paper was recycled, then greenhouse gas emissions could increase by 10% as recycling paper tends to rely more on fossil fuels than making paper from virgin fibre. But these emissions would be reduced if paper recycling is carried out using renewable energy sources rather than fossil fuels.

References

1. Agency IE (2023) CO_2 emissions in 2022. https://doi.org/10.1787/12ad1e1a-en
2. Amândio MST, Pereira JM, Rocha JMS, Serafim LS, Xavier AMRB (2022) Getting value from pulp and paper industry wastes: on the way to sustainability and circular economy. Energies 15(11). https://doi.org/10.3390/en15114105
3. Bir PM, Today R (2023) BIR: lack of recovered paper demand 'across all parts of the world', pp 1–11
4. World Paper Mill (2020). Life Cycle of Paper: The Complete Guide
5. Cordeiro J (2023) The circular economy is transforming the pulp industry. In: Valmet, Retrieved [February, 2026], https://www.valmet.com/insights/articles/experts-voice/the-circular-economy-is-transforming-the-pulp-industry/
6. The World Counts (2026) Tonnes of paper produced, in the world, this year, Retrieved [February, 2026], https://www.theworldcounts.com/challenges/other-products/environmental-impact-of-paper
7. Cozzi L, Chen O, Hyeji K (2021) The world's top 1% of emitters produce over 1000 times more CO_2 than the bottom 1% – analysis – IEA, pp 3–10. https://www.iea.org/commentaries/the-world-s-top-1-of-emitters-produce-over-1000-times-more-co2-than-the-bottom-1
8. Date P (2021) Learn the 3 key principles of the circular economy, pp 5–9
9. de Oliveira DM, de Bomfim ASC, Benini KCC, Cioffi MO, Voorwald HJ, Rodrigue D (2023) Waste paper as a valuable resource: an overview of recent trends in the polymeric composites field. Polymers 15(2):1–24. https://doi.org/10.3390/polym15020426
10. ENGIE (2020) Dashboard of Energy Transition: 2020 Edition. Published September 2020. https://www.engie.com/en/dashboard-energy-transition-2020
11. Fiksel J, Sanjay P, Raman K (2021) Steps toward a resilient circular economy in India. Clean Technol Environ Policy 23(1):203–218. https://doi.org/10.1007/s10098-020-01982-0
12. Friedlingstein P, Sullivan MO, Jones MW, Andrew RM, Bakker DCE, Hauck J, Landschützer P, Le Quéré C, Luijkx IT (2023) Global carbon budget 2023, pp 5301–5369
13. Ghosh SK (2019) Circular economy in India. Circul Econ Glob Perspect 157–185. https://doi.org/10.1007/978-981-15-1052-6_9
14. Gitarskiy ML (2019) 2019 refinement to the 2006 Ipcc guidelines for national greenhouse gas inventories. Fundam Appl Climatol 2:5–13. https://doi.org/10.21513/0207-2564-2019-2-05-13
15. Groves T (2008) reduce, reuse, recycle BMJ 336 (7650) https://doi.org/10.1136/bmj.39559.679155.47
16. Hannah R (2019) Where in the world do people emit the most CO_2? Our world in data, pp 1–5. https://ourworldindata.org/per-capita-co2
17. IBEF (2019) Industrial Development and Policy in India. IBEF Blog, India Brand Equity Foundation, pp. 1–9. www.ibef.org
18. IPCC (2014) 2013 Supplement to the 2006 IPCC Guidelines for National Greenhouse Gas Inventories: Wetlands. Hiraishi, T., Krug, T., Tanabe, K., Srivastava, N., Baasansuren, J., Fukuda, M. and Troxler, T.G. (Eds). Published: IPCC, Switzerland. pp. 7–8
19. Khadka NS (2019) Why India is a world leader in waste paper. BBC News, March 2019, pp 1–9

20. Lageman T (2020) Pulp and paper are the pioneers of the circular economy. Upmpulp, pp 1–6. https://www.upmpulp.com/media/blogs-and-stories/stories/pulp-and-paper-are-the-pioneers-of-the-circular-economy/#:~:text=Acirculareonomyisa,modelhitsclosetohome
21. Ma Z, Yang Y, Chen WQ, Wang P, Wang C, Zhang C, Gan J (2021) Material flow patterns of the global waste paper trade and potential impacts of China's import ban. Environ Sci Technol 55(13):8492–8501. https://doi.org/10.1021/acs.est.1c00642
22. Niti Aayog (2021) Govt driving transition from linear to circular economy. Government of India, Annexure, vol 1, pp 22–24. https://pib.gov.in/PressReleasePage.aspx?PRID=1705772
23. NITI Aayog (2022) Circular economy cell. National Institution for Transforming India (NITI), pp 1–4. https://niti.gov.in/circular-economy-cell
24. Odier L (2020) The 10 principles of circular economy|Lombard Odier, pp 8–12. https://www.lombardodier.com/contents/corporate-news/responsible-capital/2020/september/the-10-steps-to-a-circular-econo.html
25. Shang D, Diao G, Liu C, Yu L (2021) The impact of waste paper recycling on the carbon emissions from China's paper industry. Environ Manag 67(5):811–821. https://doi.org/10.1007/s00267-020-01417-y
26. Shiyue J, Li B, Shen Y (2021) The influence of pulp and paper industry on environment, p 02007
27. Sun Y-Y (2023) Carbon footprint factsheet. Center for Sustainable Systems, University of Michigan, Pub No. CS, pp 411–414. https://doi.org/10.4337/9781800377486.carbon.footprint
28. Tâm T, Và NCỨU, Giao CẾN, Ngh C, Chu ẢNBỤI (2016) 2006 IPCC guidelines for national greenhouse gas inventories, vol 01
29. Theodoridis T, Kraemer J (2023) Industry outlook Indian paper industry-cusp of change, pp 1–16
30. Times TP (2023) Recycled paper is no longer only a product of India or China; export decline, mills have come up in Turkey, EU, Indonesia, Vietnam, Malaysia, pp 1–12
31. Tiseo I (2023) Carbon dioxide emissions from fossil fuel combustion in India from 1960 to 2021, by type. Energy Environ 7–10. https://www.statista.com/statistics/1282696/carbon-dioxide-emissions-from-fossil-fuel-use-in-india/#:~:text=CoalcombustioninIndiaproduced,duringtheperiodinconsideration
32. van Ewijk S, Stegemann JA, Ekins P (2021) Limited climate benefits of global recycling of pulp and paper. Nat Sustain 4(2):180–187. https://doi.org/10.1038/s41893-020-00624-z
33. Van Ewijk S, Stegemann JA, Ekins P (2018) Global life cycle paper flows, recycling metrics, and material efficiency. J Ind Ecol 22(4):686–693. https://doi.org/10.1111/jiec.12613
34. What AM & Foundation EM (n.d.) Circular economy introduction. In: What is a circular economy? pp 1–7
35. World Bank (2022) CO_2 emissions from transport (% of total fuel combustion)-India 5:1–18. https://data.worldbank.org/indicator/SH.STA.DIAB.ZS?view=chart
36. World counts (2020) The world counts, pp 1–9. https://www.theworldcounts.com/search?notFound=true
37. Worldometer (2021) CO_2 emissions per capita. Worldometer, October 2021. https://www.worldometers.info/co2-emissions/co2-emissions-per-capita/

Opportunities for Green Technology Initiatives in the Indian Paper Industry

Rakesh Kumar Jain, Sunil Padhiyar, and Saakshy

Abstract Pulp and paper industry is considered among the highly resource intensive industries requiring abundant quantities of raw materials, chemicals, energy, and water. The industry continues to focus on minimizing the fresh water use and effluent discharges as a part of their move towards sustainable operating practices and closing up of the water loop. Currently, conservation and reuse of water resources have emerged as an important component of proper environmental management in the paper industry owing to the shortage of water and conflicts over water usage. The ultimate goal of the industry is to have effluent free unit operations aimed towards sustainable operation with no negative impact on process efficiency, productivity quality and environment. The traditional wastewater treatment technologies that are used in the paper mills are not able to remove recalcitrant contaminant materials; therefore, advanced wastewater treated technologies are required to either improve the wastewater biodegradability or its final quality. Thus, amid the rising water scarcity, poor internal process water quality and concern for compliance of the prescribed charter's norms, the paper industry's commitment to overall water conservation and management require the adoption of the innovative wastewater treatment strategies such as membrane filtration technology and combination of other options. The chapter discusses the status of Indian paper industry, issues and challenges on the water and wastewater management front with a focus on membrane filtration technology with a strategic approach including the process optimisation and appropriate membrane configuration to make it an environmental friendly and a green technological option for treatment of the paper mill effluent.

R. K. Jain
United Nations Industrial Development Organisation, New Delhi A-1400, India

S. Padhiyar
L D College of Engineering, Ahmedabad 380015, Gujarat, India

Saakshy (✉)
Kumarappa National Handmade Paper Institute, Jaipur 302022, Rajasthan, India
e-mail: saakshyagarwal@gmail.com

S. Sundaramurthy et al. (eds.), *Black Liquor Waste Management in the Pulp and Paper Industry*, Sustainable Materials and Technology,
https://doi.org/10.1007/978-981-96-4237-3_10

Keywords Indan paper industry · Wastewater · Effluent · Membrane filtration technology · Green technology

1 Introduction

The growth of the paper industry is intricately linked with the booming knowledge and commercial economy of India. The knowledge and commercial economy have a two-way effect on the industry. It drives the paper sector and in turn gets the required impetus of operation from the paper sector. It is well known that the core functionality of paper provides important pathways for communication, packaging or for hygienic and household uses. Further, the main reason for continued steady growth of paper consumption is its inherent properties such as low cost, light weight, flexibility, high performance and largely recyclable material. Indeed, paper can be regarded as "essential" for development and modern living. The Indian paper industry is among the fastest growing industry, ranked 15th among the top global paper producers. Despite various challenges being faced, the industry has been growing at an average growth of 6–7% against the global average of 2–3%. The industry contributes to around 4% of the global paper production. The Indian paper Industry is unique and highly fragmented, dominated by the SME's unlike the global paper industry as majority of the Indian paper mills are having smaller sizes as low as below 50 tpd. The paper industry plays a significant role in the Indian economy. According to the Pulp and Paper Times, the Indian paper and paper board packaging market was valued at $15.69 bn by 2027, registering a CAGR of 6.63% during the forecast period of 2022–2027. There are around 523 pulp and paper mills in India producing around 22 million tons of paper and paper boards with an installed capacity of around 27 million tons per annum. The paper industry provides direct employment opportunities to around 0.6 million people and indirect employment to over 1.6 million people. Writing and printing paper accounts for approximately 35% of the total paper produced. Packaging paper accounts for 55% of the total paper production in the country on account of its being eco-friendly and biodegradable [1, 2]. Packaging is also the fastest growing segment of the Indian industry. According to Indian Newsprint Manufacturers Association, the newsprint accounts for 5–7% of the paper produced in the country. Other varieties produced in the country include tissue paper/hygiene paper, insulation, filter paper, grease-proof paper, absorbent paper for laminates, etc.

Pulp and paper industry is considered among the highly energy intensive and polluting sectors. The sector largely employs conventional technologies. These produce higher quantities of effluents because they consume a lot of raw materials, chemicals, energy, and water. It is noteworthy that this industry is one of the 17 that the Central Pollution Control Board (CPCB) has classified as extremely polluting and having insufficient adherence to environmental regulations. The Ministry of Environment and Forests (MoEF), CPCB, and industry have all worked together to solve the issue of excessive concentrations of environmentally undesirable components in effluents [3]. The CPCB has stepped up its efforts to impose stricter regulations on the amount of water used per tonne of paper, the quality of treated effluent released, and energy use. Implementation of such stricter norms would be a challenge and calls for the implementation of new technologies including cleaner and greener technologies, as well as research efforts to develop cost effective solutions that could readily be adopted even by smaller units of this sector.

2 Issues and Challenges of the Indian Paper Industry

Major issues confronting the Indian pulp and paper industry are high cost of production caused by inadequate availability and non-availability of good-quality fibrous raw materials, uneconomical scale of operation, high cost of basic inputs including energy and water, technological obsolescence and environmental challenges. Despite a positive demand outlook, there are several barriers to the growth of the paper industry [4]. Some of the major issues confronting the paper industry in India are summarized in the following sections.

2.1 Raw Material Availability

Scarce availability and quality of raw material is one of the biggest challenges faced by the Indian paper industry. Unlike the developed countries, India is a fibre deficient country. The paper industry depends on a mixed source of raw materials consisting of wood, agro residue and recycled fibre. However, the supply of each source is limited, forcing the paper industry to depend on imported pulp, wood chips and waste paper. Various factors exert the influence on consistent supply of raw materials.

2.2 Cost of Basic Inputs

Rising cost of various inputs to the paper industry like fibrous raw materials, chemicals and coal severely hampered the profitability and competitive edge of the industry.

2.3 Scale of Operation and Traditional Technologies

The Indian paper industry is highly fragmented predominately consisting of small units based on recycled fibre and agro residue fibre besides few medium and large paper mills based on wood and agro residue based raw materials. Out of the around

600 number of operational paper units, around 500 paper are based on recycled waste paper as a major raw material for paper manufacturing, thereby producing approximately 15 million tons of paper and contributing to 73% of the total country's paper production. Of these, 233 paper mills are below 50 tons/day capacity, 131 units have the capacity ranging 100–300 tons/day, whereas only 21 paper units are of over 300 tons/day. There are only 22 number of wood based large paper mills with an installed capacity of more than 300 tons/day. Most of these large-scale paper mills have made substantial investments on capacity enhancements and modernization to reap the benefits of economies of scale, while the smaller scale mills still operate on traditional technologies.

2.4 Environmental Concerns

The major environmental issues faced by the paper industry include high effluent load, high color, and recycled fibre, black liquor treatment in small scale agro-based mills, solid waste treatment, reduction in air quality and growing public awareness. For compliance to strict environment regulations, the paper industry needs to invest heavily on:

- Fibre recovery system.
- Tertiary treatment systems like membrane filtration (ultra-filtration/reverse osmosis), activated carbon filter, and pressure sand filter.
- Up-gradation/modification of existing ETPs etc. to increase reuse/recycle of back liquid or treated mill effluent and achieve zero discharge level.

2.5 Lack of Skilled Manpower

Indian paper industry faces a severe shortfall of skilled manpower. There exists a big gap between requirement and availability of technically trained manpower to meet the present and the growing demand of the paper industry.

2.6 Technological Challenges

There are several technical challenges being faced by the Indian paper Industry, many of which are common across the sector and across all the three segments of the paper mills, i.e. wood, agro and RCF based. Barring few of the large wood based and some medium agro-residues and recycled fiber based medium paper mill, the industry still continues to use traditional technologies resulting in poor performance, productivity and resource efficiency besides compromising with quality and on the environmental front. Another common challenge before the industry is handling of the poor quality of process water being recycled within the various unit operation in an effort to reduce farce water consumption and wastewater discharge due to closing up of the water system.

2.7 Stickies/Slime Issues Related to Closing of the Water Systems

Closing up of the water system to reduce freshwater consumption and effluent discharge from the mills results in the build-up of slime and stickies. This issue is extremely important, and creates serious problem with quality and chemical usage issues unless the internal water management in the mills is handled correctly.

2.8 Environmental Management-Effluent Treatment

All the mills have high focus on effluent treatment. Small scale paper mills follow more or less the same regulations as larger scale mills, thus costs related to effluent treatment plant investment and energy/chemicals in effluent treatment systems are considerable. Closing up of the water system and reduced fresh water consumption has resulted in serious issues related to TDS and colour (in case of agro-/ wood-based mills) and odor issues in RCF based paper mills approaching toward zero liquid discharge.

3 Pulp and Production Process

The Indian paper industry use diversified raw materials viz. wood based, agro-residue based and recycled waste paper. Depending upon the raw materials used in the paper mills, raw material preparation and pulp mill operation may vary while the paper making process remains almost similar except the wood and agro-residue based integrated pulp and paper mill must have the appropriate black liquor management system, specifically the chemical recovery operation. Paper making process in the Indian paper industry is described in brief as follows:

Wood and agro-residues based: Raw material preparation, pulping, bleaching and chemical recovery operations are the major unit operations for the paper production in an integrated pulp and paper mill using wood and the agro-residues as the major raw materials. Pulping of raw material is required to separate the fibrous pulp and remove lignin and other impurities. Efficient removal of lignin from the raw materials improves the pulp quality, from the view point of its bleaching to improve the brightness and whiteness of the pulp.

Recycled waste paper based: Waste paper is one of the promising raw material for the production of a many varieties of paper and paper boards. The waste paper based paper mills are non-integrated and can be grouped in to:

- White bleached paper mills
- Kraft unbleached paper and paper board.

4 Green Technology-Opportunities and Potential in Indian Paper Industry

Green technologies are the technologies that have arisen during the last two decades in an effort to minimise the damage to the environment. For example: the coal fired power plants, petrochemical industries pulp and paper industry. Nevertheless, the pulp and paper industries offer several opportunities to enhance the energy efficiency. These include the upgrading of boilers, the optimization of steam systems, the recovery of waste heat, the enhancement/modernization of process control facilities, the monitoring of process equipment(s), and the adoption of energy-efficient motors and drives [5]. Though developed nations became aware of these environmental concerns in early stages of their development and accordingly made the corrective measures by adopting the environmentally benign technologies, however, developing nations like India are making its efforts and attempting to pay attention to the damaging aspects of these industries. Thus, in an effort to comply with the strict environmental legislations and to create the awareness among the masses, the Indian paper industry has taken initiatives for adoption of the alternate green and clean technology options.

Conventional technology in pulp and paper manufacturing are highly intensive in terms of raw material consumption, chemical consumption, resource consumption in the form of energy and water besides generation of higher pollution loads. The Indian paper industry is perceived to have a low rate of innovation, with a few recent developments being the improvement in the pulping and bleaching technology, increased usage of recycled paper and also the concern about the reduced consumption of fresh water and wastewater discharge and the pollution loads.. Indeed, the innovations in the industry are largely driven by environmental considerations. Wide variety of raw material usage and varied operational technologies, particularly in the Indian context, impose limits in standardization of improvements in operation.

In light of the above, the effect of industrial pollution on the environment has led the industry and research communities to focus on green technology that is concerned with developing processes and products to reduce or eliminate hazardous substances or conditions. Green technology, also called sustainable technology, is a philosophy of chemical research and engineering that encourages the design of products and processes with minimum use and generation of hazardous pollutants. Accordingly, utilization of nontoxic chemicals, environmentally benign chemicals and renewable materials are some of the key factors in green technology strategy.

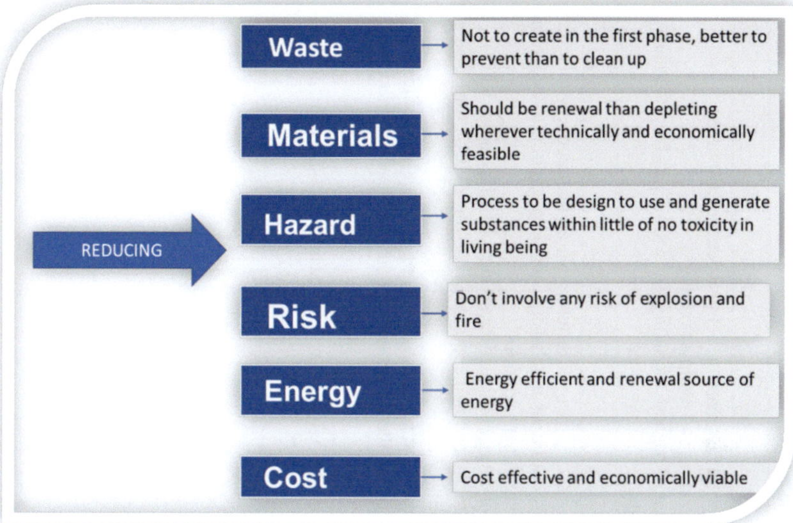

Fig. 1 Green technology concept in paper industry

Green technologies are increasingly seen as a powerful tool that utilizes a set of principles that reduces and eliminates the use of hazardous substances in the design manufacturers in application of chemical products. One of the primary goals of green technology is to prevent pollution at its source, as opposed to dealing with pollution, after it has occurred. The best form of green technology adoption in pulp and paper industry follows the concept of **REDUCING** (Fig. 1).

Various areas and process of pulp and paper manufacturing where possibility of utilization of environmentally friendly green technology initiatives is feasible for the reduction/elimination of harmful effects at source along with the opportunities for conservation of natural resources such as raw materials, chemicals, water, fuels etc., are listed in the following sections for implementation of green technologies initiatives.

5 Prevalent Technologies in the Indian Paper Industry and Potential for Adoption of Green Technologies

The potential of adoption of green technology at various stages of paper making process is given as Table 1.

Table 1 Potential of adoption of green technology at various stages [6]

Prevalent technologies in the Indian paper industry	Potential of adoption of green technology
Raw material preparation	
• Chipping of wood and bamboo-Use of traditional technology viz. disc/drum chippers and sometimes without debarking • Chopping of agro residues-traditional inefficient straw cutters • Depithing of bagasse-conventional inefficient depithing of bagasse	• De-barking of woody raw material being practiced at source • Mechanized harvesting and compact bailing • Efficient depithing of bagasse including wet washing and bio-methanation of wash spent liquor
Pulping	
• Pulping of agro residues-spherical rotary/tumbling digesters and continuous pandia digesters • Pulping of wood-generally stationary batch digesters are being used. Few mills have continuous and RDH pulping process • Generally high kappa pulping	• Separate chipping and pulping of different woody/bamboo raw materials • Use of horizontal continuous tubular digesters in place of tumbling rotatory digester for pulping of agro residues • Implementation of energy efficient Rapid Heat Displacement (RDH) pulping/super batch or codified continuous cooking (MCC) for pulping of wood based raw materials • Introduction of lower kappa pulping adopting Oxygen Delignification (ODL) process prior to the conventional/modified bleaching process
Brown stock washing	
• Poacher washers/counter current washing using rotary drum vacuum washers [1] • Chemi washers have been installed in only two mills	• Ultra-filters like double wire belt washer, twin roll washers and diffuser washers

(continued)

Table 1 (continued)

Prevalent technologies in the Indian paper industry	Potential of adoption of green technology
Bleaching	
• Hypochlorite bleaching is done in small mills [1] • Bleaching in medium and large mills is conventional-CEH, CEHH, CEoHH (C-Chlorine, E-Extraction, H-Hypochlorite, Eo-Oxidative Extraction) • Chlorine dioxide bleaching is being practiced only in few mills • Ozone bleaching is being practiced only in one mill	• Use of enzymatic pre-bleaching methods for reduction of chlorine demand during bleaching of pulps • AOX reduction in bleach plant effluent by elimination of elemental chlorine bleaching stages and by adoption of Elemental Chlorine Free (ECF) bleaching sequences • Adoption of Totally Chlorine Free (TCF) bleaching sequences for reduced environmental impact. Total Chlorine Free (TCF) process incorporates bleach enzymes
Chemical recovery	
• In evaporation street, most of the large mills have energy inefficient Long Tube Vertical (LTV) type of evaporators with the exception of two mills which still have Short Tube Vertical (STV) evaporators, followed by Direct Contact Evaporator (DCE) which are environmentally incompatible as well as it is difficult to achieve high solids concentration with DCE's, thus lower energy efficiency • Only some of the mills have installed energy efficient Falling Film (FF) type finisher effects in the multiple effect evaporator streets • In the area of recovery boiler, the mills are equipped with Tomlinson boilers with double drum technology • Use of furnace Oil in the lime kiln further results in poor energy efficiency	• Installation of free flow falling film evaporators for achieving higher steam economy during black liquor evaporation and vapor recompression evaporation for higher steam economy and low steam consumption • Thermal treatment of black liquor to achieve higher black liquor solids concentration thus higher thermal efficiency • Adoption of large capacity single drum boiler with continuous blow down • Energy efficient lime sludge re-burning-Use of producer gas and bio-gas from waste biomass in the lime kiln thus increased energy efficiency and incineration of NCG's in the lime kiln for reduced emission of sulfur compounds • Removal of non-process elements to increase the thermal efficiency of recovery boilers

(continued)

Table 1 (continued)

Prevalent technologies in the Indian paper industry	Potential of adoption of green technology
Stock preparation	
• Poor fibre recovery systems • White water recycling with poor quality of internal process water • No/little automatic control in stock preparation section [1]	• Efficient fibre recovery through sedimentation and floatation technology • Fully closed loop with minimum freshwater use • Improved automation and use of ecofriendly dyes in place of conventional dyes • Adoption of neutral and alkaline sizing agents • Installation of totally closed hoods in existing paper machines to reduce the steam consumption and recovery of heat from the hood exhaust • Installation of thermo-compressors in the paper machine to recover and reuse the flash steam • Use of ecofriendly slimicides in place of chlorine based slimicides
Paper machine	
• Slow speed machines with smaller deckle hence low production rate • Open head box without CD profile control system • Fourdrinier wire section with low retention resulting in poor paper formation • Single nip presses • Open hoods without heat recovery system • More steam consumption and low condensate recovery from dryers • No auto control system • Hardnip calendars	• High speed (avg. speed of the best paper machine in world is double in comparison to the best paper, machine in India) • Wider paper machine deckle • Efficient condensate recovery • Modification of head boxes, modification of forming section and introduction of high speed print formers [1] • Improved press configuration using nip presses and closed draw system [1]

(continued)

Table 1 (continued)

Prevalent technologies in the Indian paper industry	Potential of adoption of green technology
Environmental management	
• Effluent treatment employing conventional ASP process-Difficult to achieve desired pollution norms • Inefficient control of gaseous emission (NCG's)	• Bio-remediation of color, COD & AOX from paper mill effluent • Efficient recycling and re-use of water, application of filtration based technologies (ultra and nanofiltration) for the treatment of the effluents for TDS and color removal • Collection and incineration of Non-Condensable Gases (NCGs) to control the odor from pulp mill and the evaporator plant • Use of energy efficient pumps and motors to reduce the power consumption in pulping and paper making operations
Recycled fibre-Bio deinking	
• High consumption of deinking and bleach chemicals and still lower toner ink removal • Poor drainage rate resulting in lower productivity • Inferior product quality • Lower brightness • Higher stickies • Higher dirt count • High ERIC values • Higher pollution loads due to high BOD and COD values	• Savings in deinking and bleach chemicals • Improved drainage rate thereby enhanced productivity • Improved product quality ◦ Gain in brightness ◦ Reduction in stickies ◦ Reduction in dirt count ◦ Low ERIC values • Reduced pollution load due to low BOD and COD value
Enzymatic refining of pulp	
• Not used much in practice	• Reduction in refining energy • Improved drainability

6 Environmental Issues and Challenges in the Indian Paper Industry

The Indian paper industry is highly water intensive, consuming large quantity of water. As a large water consumer, the industry has been progressively improving its water usages while achieving stipulated norms, discharges huge quantities of polluted effluents whose correct treatment prior to discharge into the environment is critical.

Faced with issues and challenges, one of the major challenges is on the environmental front, which the industry has been able to address gradually by adopting various measures such as adoption of cleaner production technologies attempting towards source reduction of pollutants coupled with end of pipe treatment measures. Closed water circuit is being practiced in majority of the paper mills as measures for minimization of freshwater consumption owing to the lack of water resources and increasing conflicts over the water usages. This has negative impact on the product quality, productivity and the overall process efficiency of the paper mill.

Currently, conservation and reuse of water resources have emerged as an important component of proper environmental management owing to the shortage of water and conflicts over water usage [7]. There is a worldwide tendency towards development and application of advanced wastewater treatment systems that enable significant improvements in the final wastewater quality by applying standards that are increasingly becoming restrictive or enabling the reuse of treated wastewater in the production process itself or in other beneficial uses rather than discharging it into water bodies. In the pulp and paper industry, the concept of water circuits that are either entirely or partially closed has settled in [8]. The driving force for freshwater minimization might be more efficient processes, lack of water resources, fresh water and wastewater costs as well as stringent norms, but today all agree that it has to be done.

Though a lot has been done during the last decade and many of the RCF based paper mills claims to operate with specific water consumption as low as less than 10 m3/tpd, however, these mills are faced with various issues and challenges related to poor quality of the internal process water, build-up of TDS, TSS and microbial slime, foul smell and stickies, with negative impact on product quality, productivity and overall process efficiency of the paper mills. Pertaining to the quality of the process water being recycled in the mill, it is not clear what are the tolerance limits for the water quality in different mills. Must the reused water really be potable? or could the reused water contain "something" without harming the product or the viability of the process? Does that something differ from mill to mill? Based on the knowledge today, it seems that no measured value commensurately describes the paper making process in such a way that common conclusion could be drawn.

6.1 Internal Process Water Characteristics

Poor quality of the internal process water has been as one of the major challenges. This is due to the build-up of the organic and inorganics contents coupled with the issue related to stickies and slimes. The paper mill water carries significant quantities of fibers (losses with effluent 0.5–5% of the total fiber amount), filler, fines and other wet-end additives that contribute to TSS, COD, and BOD. TSS varies significantly from mill to mill based on the type of internal clarification equipment used and equipment design and its arrangement. COD depends on the amount of TSS and other chemically oxidizable wet-end additives such as starch. High BOD due to the presence of large amount of biologically oxidizable material including the above mentioned and others such as dry strength and wet strength resins, drainage aids, sizing chemicals and other dissolved organics.

In general, most of the Indian pulp and paper mills have effluent treatment plants based on activated sludge process for the treatment of mill effluent. The process involves primary clarifier (for reduction in suspended solids) followed by aeration tank containing active biomass or Mixed Liquor Suspended Solids (MLSS) and surface/diffused aerators for the oxidation of biodegradable organic matter. The treated effluent after aeration tank is subjected to secondary clarifier before being discharged into the local drain/recipient stream. In recent times, some mills have incorporated equalisation tank before the primary clarifier to arrest the fluctuation in pollution load due to variation in raw material/product quality or shock load and consequently its impact on the ETP performance. For further polishing the quality of effluent, some mills have adopted tertiary treatment system involving physico—chemical treatment system like clarifloculators, pressure sand filter, dual media filter etc.

6.2 Internal Process Water and Its Handling

The internal water handling in majority of the mills appears not to the mark as the awareness about its handling and treatment in general is low. There is a good possibility of improving/managing their paper quality by following best available technologies/practices of the internal water handling. For a better control of internal water handling, the better understanding of the different wastewater streams/water circuits would help in the reduction of consumption of fresh water and pollution loads in discharged effluents with added advantages of fibre recovery. The figure shows that the water system of a paper mill may consist of three water loops: the primary circuits, the secondary circuits and in some cases the tertiary circuits (Fig. 2).

Over the past decades, the pulp and paper industry have continued to focus on minimizing the fresh water use and effluent discharges as a part of their move towards sustainable operating practices [9]. Three stages-basic conservation, water reuse and water recycling-provide a systematic approach with optimal utilization of individual

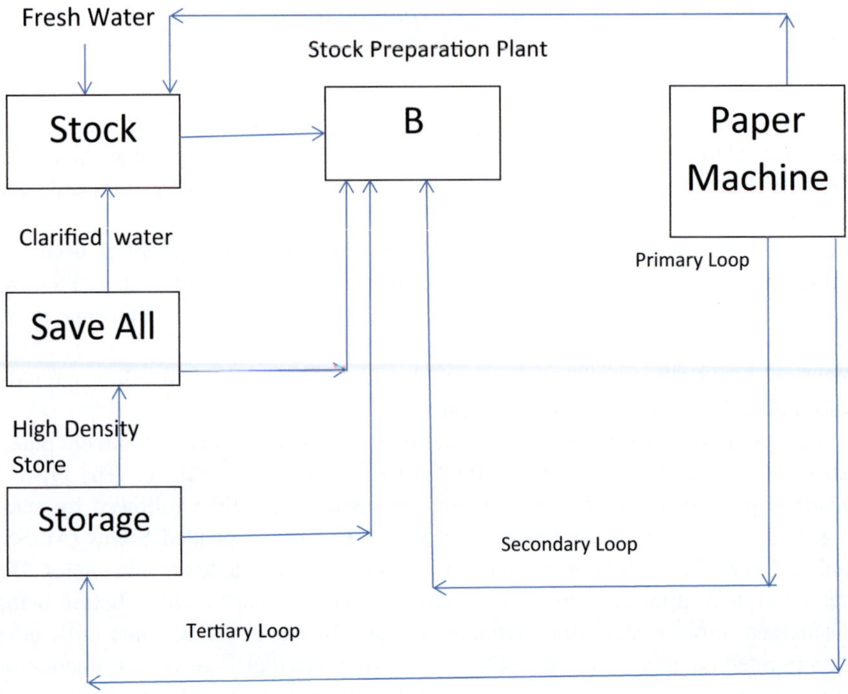

Fig. 2 Water system of a paper mill

wastewater streams, good housekeeping, and segregation of wastewater streams. The ultimate goal of the industry is to have effluent free units with no negative environmental impact.

A lot of work has been done during the last decade and many mills are already running with specific water consumption below the prescribed standard limits. However, these mills are struggling with various challenges and issues related to the quality, productivity and overall process efficiency because of the poor quality of the internal process water caused due to closing up of the water circuits. Besides, the quality of the internal process water specifically in the recycled fibre based paper mills is found to be highly deteriorated due the presence of microbial slime and stickies.

The choice of method for internal water treatment needs to be done based on the quality requirement of the treated water. Practically, it is necessary to know which components needs to be removed (suspended solids, colloids) to make the treated water reusable (Table 2).

Table 2 Type of wastewater generated from various paper making process

Raw material preparation section	For cleaning and washing of raw material
Pulping section	Maintenance of bath ratio and preparation of pulping chemicals (white liquor)
Pulp washing and bleaching	Pulp washers of final stage and maintenance of consistency in various stages of bleaching as well as bleach chemical preparation, cleaning of washers
Stock preparation	Dilution of pulp stock as well as preparation of additives and chemicals
Paper machine	Paper machine showers for cleaning of the wire and belts
Utilities	Power and chemical recovery boiler, vacuum pump sealing, gland cooling etc.
Miscellaneous	Horticulture, drinking, domestic use (in case of residential colonies within the mill campus premises)

6.3 Wastewater Management Practices in the Indian Paper Industry

The water requirement in pulp and paper mills depends mainly on raw material usage, process and technology employed and the quality of end product. In this perspective, the Indian paper industry, (mainly the medium and small mills) has been traditionally viewed to have significant water and energy footprints due to use of mixed raw materials, production of multiple grades of paper, use of multiple paper machine, traditional technology and process equipments. In the last decade, majority of the large wood based and medium size agro and recycled waste paper based mills have taken adequate measures to contain pollution by way of upgrading their technologies in a phased manner for at-source reduction of pollutants and consumption of fresh water and pollutants loads followed by adopting advanced effluent treatment technologies as an end-of-pipe treatment option.

6.4 Closing the Water Loops

Recent technological developments in paper mill have resulted in a decrease in water consumption and specific emissions by adopting closed water loops and internal process wastewater treatment. Depending on the techniques applied, there are operational problem and relatively high cost involved. However, these techniques can be considered seriously during the upgrading of existing mills or building new mills. Techniques for water reclamation can also be used to reduce the fresh water consumption in existing mills without having the intention of completely closing the water system. The ideal approach should be at source reduction of pollution reduction and waste generation rather than the end-of-the pipe treatment approach. Until now, closed water loops have been realized in recycled fibre based paper mills. In Germany, about 10 paper mills producing corrugating medium and test liner paper have closed water circuits [10].

6.5 Internal Process Water Treatment/Purification

Segregation provides the possibility to treat process waters or effluents individually, close to where they are generated. With segregation of clean and contaminated water fractions, a significant reduction of effluent volume can be achieved. It is not known what the tolerance limits are for water quality in different mills. Must the reused water really be potable? or could the reused water contain "something" without harming the product or the viability of the process? [7]. The traditional wastewater treatment technologies that are used in the paper mills are not able to remove recalcitrant contaminant materials; therefore, advanced wastewater treatment technologies are required to either improve the wastewater biodegradability or its final quality [11]. Membrane filtration has gradually becoming a more versatile treatment option providing possibilities for recycling and reuse of mill water with benefits' of high level of contaminant removal, less space requirement and low energy consumption.

6.6 Membrane Filtration Technologies in the Pulp and Paper Industry

Membrane treatment in the pulp and paper industry may serve to optimize the loop closer and therefore, helps to reduce fresh water intake as well as wastewater treatment. Other useful purposes of the process are: reduction and removal of non-biodegradable organic/inorganic substances, improved product quality because of lower pollution of loop water, reuse of treated effluent in the production, recovery of valuable substances such as lignin based products, coating pigments and minimizing environmental impact because of improved effluent quality.

Membrane technology is being adopted increasingly where manufacturers are closing their mill production process. The membranes have proved to be a viable technology for treating both TCF (total chlorine free) and ECF (elemental chlorine free) bleaching streams. For TCF bleaching, membranes also provide an effective method of closing the chelating stages and allow the filtrates to be used as bleach plant wash liquors [12–14]. Raw water use represents a significant cost. Paper making machinery generates large quantities of white water, which can be recycled, thereby achieving major savings on water usage. Any impurities in the recycling stream could have a disastrous effect on the product quality. Membrane filtration offers a physical barrier to impurities and when combined with the wide range of membrane types, it allows efficient recycling of water. Membrane technology offers a practical process for the purification of macromolecules species in aqueous solutions. The application in the pulp and paper industry at present pose several conditions for the membranes and many researchers are investigating the feasibility. The need to reduce water consumption and reuse treated effluent has particularly catalyzed effects to find new targets for membrane separation techniques such as the use of ultrafiltration of debarking and bleaching effluents.

Membrane filtration can be divided into four broad groups, each determined by the size of particle which can be retained by the membrane material. These range from reverse osmosis which provides the finest level of filtration, through nanofiltration and ultrafiltration to microfiltration which uses the course of membranes. Ultrafiltration, which can separate particles up to a few tenths of a micron in diameter of different molecular weights, is widely used in the forest products industry. Membrane filtration offers several benefits to the pulp and paper industry. It has been used to remove COD, colour, heavy metals and total dissolved solids from the pulp and paper mill effluents.

7 Conclusions

The development, promotion and adoption of green technologies which increasingly use renewable resources; reduce wastes, pollutants, emissions; reduce the pressure on natural resources should help to restore the balance of the eco-system and biosphere and ultimately help in providing "Ecologically Sustainable Development" [7]. Amid the rising water scarcity, poor internal process water quality and concern for compliance of the prescribed charter's norms, the Indian paper industry's commitment to overall water conservation and management require adoption of the innovative water treatment strategies such as membrane filtration technology and combination of other options. Membrane filtration technology, by adopting a strategic approach including the process optimisation and appropriate membrane configuration, should likely become a techno-economically and environmental friendly green technology option for the treatment of the paper mill effluent to reduce fresh water consumption and the effluent discharge to comply with the specified norms.

References

1. GIZ (2021) Indian Pulp & Paper Sector - An insight for an energy efficient tomorrow, A study by the Centre Pulp & Paper Research Institute, India, Retrieved [February, 2026], https://demo.tuewas-asia.org/wp-content/uploads/2021/10/Pulp-Paper-Sector-Study.pdf
2. Malik V, Sharma M, Bandyopadhyay S (2023) Indian paper industry-cusp of change, pp 1–16
3. Kanth GVR, Suri RN (2014) Green technologies. Office of the Technology Missions, IISC, Bangalore. http://Icm.csa.iisc.ernet.in/TDM/green.html
4. Thapliyal BP (2020) Status of paper industry, Powerpoint presentation, Retrieved [February, 2026], https://www.scribd.com/presentation/464818501/Dr-B-P-Thapliyal-Director-CPPRI

5. Thapliyal BP (2024) An overview of best energy management practices in pulp and paper-making processes for improving energy efficiency in the pulp and paper sector. In: Sectoral Workshop on Best Practices in Energy Efficiency in Pulp & paper Sector, Amritsar, India
6. Jain RK (2015) Compendium of census survey of Indian paper industry. Central Pulp & Paper Research Institute, Saharanpur, India, pp 107–109
7. Moving from Trade-offs to Win-Wins in Water Management (2019) PaperMart, pp 1–29
8. Nuortila-Jokinen J (1986) Fresh water minimization by membrane filtration in the pulp and paper industry. Encycl Life Supp Syst II (2016)
9. Holik H (2006). Hand book of paper and paper board. Wiley-VCH Verlag Gmbh & Co. KGaA, Weinheim. ISBN-13 978-527-30997-9
10. Wardencki W, Curylo J, Namiesnik J (2005) Green chemistry–current and future issues. Pol J Environ Stud 14(4):389–395
11. Kersten A, Putz HJ, Schabel S (2013) Mineral oil in board and paper recycling. In: Handbook of environmental chemistry, vol 23, pp 399–417. https://doi.org/10.1007/698_2012_183
12. Hatzinger PB, Whittier MC, Arkins MD, Bryan CW, Guarini WJ (2002) In-situ and ex-situ bioremediation options for treating perchlorate in groundwater. Remediation 12(2):69–86. https://doi.org/10.1002/rem.10026
13. Jain RK, Kumar V, Anders I (2018) Diagnostic assessment & strategic approach adopting advanced technologies for water management in Indian paper industry. In: National conference on pulp, paper & allied industries, Coimbatore, India
14. Jain RK, Oyvind E, Ander I, Kumar V (2017) Development and adoption of appropriate technologies for enhancing sustainability and competitiveness of the Indian paper industry. In: A research paper in proceedings of the 13th international technical conference on pulp, paper and allies industries, New Delhi, India, pp 41–52